Student Edition

Eureka Math
Grade 6
Modules 1 & 2

Special thanks go to the Gordon A. Cain Center and to the Department of Mathematics at Louisiana State University for their support in the development of *Eureka Math*.

For a free *Eureka Math* Teacher Resource Pack, Parent Tip Sheets, and more please visit www.Eureka.tools

Published by the non-profit Great Minds

Copyright © 2015 Great Minds. No part of this work may be reproduced, sold, or commercialized, in whole or in part, without written permission from Great Minds. Non-commercial use is licensed pursuant to a Creative Commons Attribution-NonCommercial-ShareAlike 4.0 license; for more information, go to http://greatminds.net/maps/math/copyright. "Great Minds" and "Eureka Math" are registered trademarks of Great Minds.

Printed in the U.S.A.

This book may be purchased from the publisher at eureka-math.org

10 9 8 7 6 5 4 3 2

ISBN 978-1-63255-312-6

Lesson 1: Ratios

Classwork

Example 1

The coed soccer team has four times as many boys on it as it has girls. We say the ratio of the number of boys to the number of girls on the team is $4: 1$. We read this as *four to one*.

Suppose the ratio of the number of boys to the number of girls on the team is $3: 2$.

Example 2: Class Ratios

Write the ratio of the number of boys to the number of girls in our class.

Write the ratio of the number of girls to the number of boys in our class.

Lesson 1:   Ratios

S.1

©2015 Great Minds. eureka-math.org
G6-M1-SE-B1-1.3.1-01.2016

Record a ratio for each of the examples the teacher provides.

1. _____   2. _____

3. _____   4. _____

5. _____   6. _____

Exercise 1

My own ratio compares _____ to

_____ .

My ratio is _____ .

Exercise 2

Using words, describe a ratio that represents each ratio below.

a. 1 to 12 _____

_____ .

b. 12 : 1 _____

_____ .

c. 2 to 5 _____

_____ .

2015 Great Minds. eureka-math.org
G6-M1-SE-B1-1.3.1-01.2016

EUREKA
MATH™

d.   5 to 2   _____

_____.

e.   10: 2   _____

_____.

f.   2: 10   _____

_____.

©2015 Great Minds. eureka-math.org
G6-M1-SE-B1-1.3.1-01.2016

Lesson Summary

A ratio is an ordered pair of numbers, which are not both zero.

A ratio is denoted $A : B$ to indicate the order of the numbers—the number A is first, and the number B is second.

The order of the numbers is important to the meaning of the ratio. Switching the numbers changes the relationship. The description of the ratio relationship tells us the correct order for the numbers in the ratio.

Problem Set

1. At the sixth-grade school dance, there are 132 boys, 89 girls, and 14 adults.
   a. Write the ratio of the number of boys to the number of girls.
   b. Write the same ratio using another form ($A : B$ vs. A to B).
   c. Write the ratio of the number of boys to the number of adults.
   d. Write the same ratio using another form.

2. In the cafeteria, 100 milk cartons were put out for breakfast. At the end of breakfast, 27 remained.
   a. What is the ratio of the number of milk cartons taken to the total number of milk cartons?
   b. What is the ratio of the number of milk cartons remaining to the number of milk cartons taken?

3. Choose a situation that could be described by the following ratios, and write a sentence to describe the ratio in the context of the situation you chose.

   For example:

   $3 : 2$. When making pink paint, the art teacher uses the ratio $3 : 2$. For every 3 cups of white paint she uses in the mixture, she needs to use 2 cups of red paint.
   a. 1 to 2
   b. 29 to 30
   c. $52 : 12$

2015 Great Minds. eureka-math.org
G6-M1-SE-B1-1.3.1-01.2016

Lesson 2: Ratios

Classwork

Exercise 1

Come up with two examples of ratio relationships that are interesting to you.

1.

2.

Exploratory Challenge

A T-shirt manufacturing company surveyed teenage girls on their favorite T-shirt color to guide the company's decisions about how many of each color T-shirt they should design and manufacture. The results of the survey are shown here.

Favorite T-Shirt Colors of Teenage Girls Surveyed

			X			
			X			
			X	X		
	X		X	X		X
	X		X	X	X	X
	X	X	X	X	X	X
X	X	X	X	X	X	X
Red	Blue	Green	White	Pink	Orange	Yellow

Exercises for Exploratory Challenge

1. Describe a ratio relationship, in the context of this survey, for which the ratio is 3: 5.

2. For each ratio relationship given, fill in the ratio it is describing.

Description of the Ratio Relationship (Underline or highlight the words or phrases that indicate the description is a ratio.)	Ratio
For every 7 white T-shirts they manufacture, they should manufacture 4 yellow T-shirts. The ratio of the number of white T-shirts to the number of yellow T-shirts should be …	
For every 4 yellow T-shirts they manufacture, they should manufacture 7 white T-shirts. The ratio of the number of yellow T-shirts to the number of white T-shirts should be …	
The ratio of the number of girls who liked a white T-shirt best to the number of girls who liked a colored T-shirt best was …	
For each red T-shirt they manufacture, they should manufacture 4 blue T-shirts. The ratio of the number of red T-shirts to the number of blue T-shirts should be …	
They should purchase 4 bolts of yellow fabric for every 3 bolts of orange fabric. The ratio of the number of bolts of yellow fabric to the number of bolts of orange fabric should be …	
The ratio of the number of girls who chose blue or green as their favorite to the number of girls who chose pink or red as their favorite was …	
Three out of every 26 T-shirts they manufacture should be orange. The ratio of the number of orange T-shirts to the total number of T-shirts should be …	

3. For each ratio given, fill in a description of the ratio relationship it could describe, using the context of the survey.

Description of the Ratio Relationship (Underline or highlight the words or phrases that indicate your example is a ratio.)	Ratio
	4 to 3
	3 : 4
	19 : 7
	7 to 26

Lesson 2: Ratios

EUREKA MATH™

2015 Great Minds. eureka-math.org
G6-M1-SE-B1-1.3.1-01.2016

Lesson Summary

- Ratios can be written in two ways: A to B or $A : B$.
- We describe ratio relationships with words, such as *to*, *for each*, *for every*.
- The ratio $A : B$ is not the same as the ratio $B : A$ (unless A is equal to B).

Problem Set

1. Using the floor tiles design shown below, create 4 different ratios related to the image. Describe the ratio relationship, and write the ratio in the form $A : B$ or the form A to B.

2. Billy wanted to write a ratio of the number of apples to the number of peppers in his refrigerator. He wrote $1 : 3$. Did Billy write the ratio correctly? Explain your answer.

This page intentionally left blank

Lesson 3: Equivalent Ratios

Classwork

Exercise 1

Write a one-sentence story problem about a ratio.

Write the ratio in two different forms.

Exercise 2

Shanni and Mel are using ribbon to decorate a project in their art class. The ratio of the length of Shanni's ribbon to the length of Mel's ribbon is $7:3$.

Draw a tape diagram to represent this ratio.

Exercise 3

Mason and Laney ran laps to train for the long-distance running team. The ratio of the number of laps Mason ran to the number of laps Laney ran was 2 to 3.

    a.   If Mason ran 4 miles, how far did Laney run? Draw a tape diagram to demonstrate how you found the answer.

    b.   If Laney ran 930 meters, how far did Mason run? Draw a tape diagram to determine how you found the answer.

    c.   What ratios can we say are equivalent to $2:3$?

2015 Great Minds. eureka-math.org
6-M1-SE-B1-1.3.1-01.2016

Exercise 4

Josie took a long multiple-choice, end-of-year vocabulary test. The ratio of the number of problems Josie got incorrect to the number of problems she got correct is 2: 9.

    a.   If Josie missed 8 questions, how many did she get correct? Draw a tape diagram to demonstrate how you found the answer.

    b.   If Josie missed 20 questions, how many did she get correct? Draw a tape diagram to demonstrate how you found the answer.

    c.   What ratios can we say are equivalent to 2: 9?

©2015 Great Minds. eureka-math.org
G6-M1-SE-B1-1.3.1-01.2016

d. Come up with another possible ratio of the number Josie got incorrect to the number she got correct.

e. How did you find the numbers?

f. Describe how to create equivalent ratios.

2015 Great Minds. eureka-math.org
6-M1-SE-B1-1.3.1-01.2016

Lesson Summary

Two ratios $A:B$ and $C:D$ are *equivalent ratios* if there is a nonzero number c such that $C = cA$ and $D = cB$. For example, two ratios are equivalent if they both have values that are equal.

Ratios are equivalent if there is a nonzero number that can be multiplied by both quantities in one ratio to equal the corresponding quantities in the second ratio.

Problem Set

1. Write two ratios that are equivalent to $1:1$.

2. Write two ratios that are equivalent to $3:11$.

3.
    a. The ratio of the width of the rectangle to the height of the rectangle is _____ to _____.

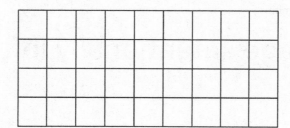

    b. If each square in the grid has a side length of 8 mm, what is the width and height of the rectangle?

4. For a project in their health class, Jasmine and Brenda recorded the amount of milk they drank every day. Jasmine drank 2 pints of milk each day, and Brenda drank 3 pints of milk each day.

    a. Write a ratio of the number of pints of milk Jasmine drank to the number of pints of milk Brenda drank each day.

    b. Represent this scenario with tape diagrams.

    c. If one pint of milk is equivalent to 2 cups of milk, how many cups of milk did Jasmine and Brenda each drink? How do you know?

    d. Write a ratio of the number of cups of milk Jasmine drank to the number of cups of milk Brenda drank.

    e. Are the two ratios you determined equivalent? Explain why or why not.

©2015 Great Minds. eureka-math.org
G6-M1-SE-B1-1.3.1-01.2016

This page intentionally left blank

Lesson 4: Equivalent Ratios

Classwork

Example 1

The morning announcements said that two out of every seven sixth-grade students in the school have an overdue library book. Jasmine said, "That would mean 24 of us have overdue books!" Grace argued, "No way. That is way too high." How can you determine who is right?

Exercise 1

Decide whether or not each of the following pairs of ratios is equivalent.

- If the ratios are not equivalent, find a ratio that is equivalent to the first ratio.
- If the ratios are equivalent, identify the nonzero number, c, that could be used to multiply each number of the first ratio by in order to get the numbers for the second ratio.

a. $6\!:\!11$ and $42\!:\!88$

_____ Yes, the value, c, is _____.

_____ No, an equivalent ratio would be _____.

b. $0\!:\!5$ and $0\!:\!20$

_____ Yes, the value, c, is _____.

_____ No, an equivalent ratio would be _____.

©2015 Great Minds. eureka-math.org
G6-M1-SE-B1-1.3.1-01.2016

Exercise 2

In a bag of mixed walnuts and cashews, the ratio of the number of walnuts to the number of cashews is $5:6$. Determine the number of walnuts that are in the bag if there are 54 cashews. Use a tape diagram to support your work. Justify your answer by showing that the new ratio you created of the number of walnuts to the number of cashews is equivalent to $5:6$.

2015 Great Minds. eureka-math.org
6-M1-SE-B1-1.3.1-01.2016

Lesson Summary

Recall the description:

Two ratios $A: B$ and $C: D$ are *equivalent ratios* if there is a positive number, c, such that $C = cA$ and $D = cB$. For example, two ratios are equivalent if they both have values that are equal.

Ratios are equivalent if there is a positive number that can be multiplied by both quantities in one ratio to equal the corresponding quantities in the second ratio.

This description can be used to determine whether two ratios are equivalent.

Problem Set

1. Use diagrams or the description of equivalent ratios to show that the ratios 2: 3, 4: 6, and 8: 12 are equivalent.

2. Prove that 3: 8 is equivalent to 12: 32.
    a. Use diagrams to support your answer.
    b. Use the description of equivalent ratios to support your answer.

3. The ratio of Isabella's money to Shane's money is 3: 11. If Isabella has $33, how much money do Shane and Isabella have together? Use diagrams to illustrate your answer.

This page intentionally left blank

Lesson 5: Solving Problems by Finding Equivalent Ratios

Classwork

Example 1

A County Superintendent of Highways is interested in the numbers of different types of vehicles that regularly travel within his county. In the month of August, a total of 192 registrations were purchased for passenger cars and pickup trucks at the local Department of Motor Vehicles (DMV). The DMV reported that in the month of August, for every 5 passenger cars registered, there were 7 pickup trucks registered. How many of each type of vehicle were registered in the county in the month of August?

    a.   Using the information in the problem, write four different ratios and describe the meaning of each.

    b.   Make a tape diagram that represents the quantities in the part-to-part ratios that you wrote.

    c.   How many equal-sized parts does the tape diagram consist of?

    d.   What total quantity does the tape diagram represent?

©2015 Great Minds. eureka-math.org
G6-M1-SE-B1-1.3.1-01.2016

e. What value does each individual part of the tape diagram represent?

f. How many of each type of vehicle were registered in August?

Example 2

The Superintendent of Highways is further interested in the numbers of commercial vehicles that frequently use the county's highways. He obtains information from the Department of Motor Vehicles for the month of September and finds that for every 14 non-commercial vehicles, there were 5 commercial vehicles. If there were 108 more non-commercial vehicles than commercial vehicles, how many of each type of vehicle frequently use the county's highways during the month of September?

2015 Great Minds. eureka-math.org
6-M1-SE-B1-1.3.1-01.2016

EUREKA MATH™

Exercises

1. The ratio of the number of people who own a smartphone to the number of people who own a flip phone is 4: 3. If 500 more people own a smartphone than a flip phone, how many people own each type of phone?

2. Sammy and David were selling water bottles to raise money for new football uniforms. Sammy sold 5 water bottles for every 3 water bottles David sold. Together they sold 160 water bottles. How many did each boy sell?

3. Ms. Johnson and Ms. Siple were folding report cards to send home to parents. The ratio of the number of report cards Ms. Johnson folded to the number of report cards Ms. Siple folded is 2: 3. At the end of the day, Ms. Johnson and Ms. Siple folded a total of 300 report cards. How many did each person fold?

4. At a country concert, the ratio of the number of boys to the number of girls is 2: 7. If there are 250 more girls than boys, how many boys are at the concert?

Problem Set

1. Last summer, at *Camp Okey-Fun-Okey*, the ratio of the number of boy campers to the number of girl campers was 8 : 7. If there were a total of 195 campers, how many boy campers were there? How many girl campers?

2. The student-to-faculty ratio at a small college is 17 : 3. The total number of students and faculty is 740. How many faculty members are there at the college? How many students?

3. The Speedy Fast Ski Resort has started to keep track of the number of skiers and snowboarders who bought season passes. The ratio of the number of skiers who bought season passes to the number of snowboarders who bought season passes is 1 : 2. If 1,250 more snowboarders bought season passes than skiers, how many snowboarders and how many skiers bought season passes?

4. The ratio of the number of adults to the number of students at the prom has to be 1 : 10. Last year there were 477 more students than adults at the prom. If the school is expecting the same attendance this year, how many adults have to attend the prom?

2015 Great Minds. eureka-math.org
6-M1-SE-B1-1.3.1-01.2016

Lesson 6: Solving Problems by Finding Equivalent Ratios

Classwork

Exercises

1. The Business Direct Hotel caters to people who travel for different types of business trips. On Saturday night there is not a lot of business travel, so the ratio of the number of occupied rooms to the number of unoccupied rooms is $2:5$. However, on Sunday night the ratio of the number of occupied rooms to the number of unoccupied rooms is $6:1$ due to the number of business people attending a large conference in the area. If the Business Direct Hotel has 432 occupied rooms on Sunday night, how many unoccupied rooms does it have on Saturday night?

2. Peter is trying to work out by completing sit-ups and push-ups in order to gain muscle mass. Originally, Peter was completing five sit-ups for every three push-ups, but then he injured his shoulder. After the injury, Peter completed the same number of repetitions as he did before his injury, but he completed seven sit-ups for every one push-up. During a training session after his injury, Peter completed eight push-ups. How many push-ups was Peter completing before his injury?

©2015 Great Minds. eureka-math.org
G6-M1-SE-B1-1.3.1-01.2016

3. Tom and Rob are brothers who like to make bets about the outcomes of different contests between them. Before the last bet, the ratio of the amount of Tom's money to the amount of Rob's money was $4:7$. Rob lost the latest competition, and now the ratio of the amount of Tom's money to the amount of Rob's money is $8:3$. If Rob had $280 before the last competition, how much does Rob have now that he lost the bet?

4. A sporting goods store ordered new bikes and scooters. For every 3 bikes ordered, 4 scooters were ordered. However, bikes were way more popular than scooters, so the store changed its next order. The new ratio of the number of bikes ordered to the number of scooters ordered was $5:2$. If the same amount of sporting equipment was ordered in both orders and 64 scooters were ordered originally, how many bikes were ordered as part of the new order?

5. At the beginning of Grade 6, the ratio of the number of advanced math students to the number of regular math students was $3:8$. However, after taking placement tests, students were moved around changing the ratio of the number of advanced math students to the number of regular math students to $4:7$. How many students started in regular math and advanced math if there were 92 students in advanced math after the placement tests?

EUREKA
MATH™

2015 Great Minds. eureka-math.org
G6-M1-SE-B1-1.3.1-01.2016

6. During first semester, the ratio of the number of students in art class to the number of students in gym class was 2: 7. However, the art classes were really small, and the gym classes were large, so the principal changed students' classes for second semester. In second semester, the ratio of the number of students in art class to the number of students in gym class was 5: 4. If 75 students were in art class second semester, how many were in art class and gym class first semester?

7. Jeanette wants to save money, but she has not been good at it in the past. The ratio of the amount of money in Jeanette's savings account to the amount of money in her checking account was 1: 6. Because Jeanette is trying to get better at saving money, she moves some money out of her checking account and into her savings account. Now, the ratio of the amount of money in her savings account to the amount of money in her checking account is 4: 3. If Jeanette had $936 in her checking account before moving money, how much money does Jeanette have in each account after moving money?

Lesson Summary

When solving problems in which a ratio between two quantities changes, it is helpful to draw a *before* tape diagram and an *after* tape diagram.

Problem Set

1. Shelley compared the number of oak trees to the number of maple trees as part of a study about hardwood trees in a woodlot. She counted 9 maple trees to every 5 oak trees. Later in the year there was a bug problem and many trees died. New trees were planted to make sure there was the same number of trees as before the bug problem. The new ratio of the number of maple trees to the number of oak trees is $3:11$. After planting new trees, there were 132 oak trees. How many more maple trees were in the woodlot before the bug problem than after the bug problem? Explain.

2. The school band is comprised of middle school students and high school students, but it always has the same maximum capacity. Last year the ratio of the number of middle school students to the number of high school students was $1:8$. However, this year the ratio of the number of middle school students to the number of high school students changed to $2:7$. If there are 18 middle school students in the band this year, how many fewer high school students are in the band this year compared to last year? Explain.

EUREKA
MATH™

2015 Great Minds. eureka-math.org
G6-M1-SE-B1-1.3.1-01.2016

Lesson 7: Associated Ratios and the Value of a Ratio

Classwork

Example 1

Which of the following correctly models that the number of red gumballs is $\frac{5}{3}$ the number of white gumballs?

a. Red   ▢▢▢
    White ▢▢▢▢▢

b. Red   ▢▢▢▢▢
    White ▢▢▢

c. Red   ▢▢▢
    White ▢▢

d. Red   ▢▢▢▢
    White ▢▢▢▢▢▢▢

Example 2

The duration of two films are modeled below.

Film A ▢▢▢▢▢

Film B ▢▢▢▢▢▢▢

a. The ratio of the length of Film A to the length of Film B is _____ : _____.

b. The length of Film A is $\dfrac{\Box}{\Box}$ of the length of Film B.

c. The length of Film B is $\dfrac{\Box}{\Box}$ of the length of Film A.

EUREKA
MATH

Lesson 7:   Associated Ratios and the Value of a Ratio

S.27

©2015 Great Minds. eureka-math.org
G6-M1-SE-B1-1.3.1-01.2016

Exercise 1

Sammy and Kaden went fishing using live shrimp as bait. Sammy brought 8 more shrimp than Kaden brought. When they combined their shrimp they had 32 shrimp altogether.

   a.   How many shrimp did each boy bring?

   b.   What is the ratio of the number of shrimp Sammy brought to the number of shrimp Kaden brought?

   c.   Express the number of shrimp Sammy brought as a fraction of the number of shrimp Kaden brought.

   d.   What is the ratio of the number of shrimp Sammy brought to the total number of shrimp?

   e.   What fraction of the total shrimp did Sammy bring?

2015 Great Minds. eureka-math.org
G6-M1-SE-B1-1.3.1-01.2016

Exercise 2

A food company that produces peanut butter decides to try out a new version of its peanut butter that is extra crunchy, using twice the number of peanut chunks as normal. The company hosts a sampling of its new product at grocery stores and finds that 5 out of every 9 customers prefer the new extra crunchy version.

    a.   Let's make a list of ratios that might be relevant for this situation.

        i.   The ratio of number preferring new extra crunchy to total number surveyed is _____.

        ii.   The ratio of number preferring regular crunchy to the total number surveyed is _____.

        iii.   The ratio of number preferring regular crunchy to number preferring new extra crunchy is _____.

        iv.   The ratio of number preferring new extra crunchy to number preferring regular crunchy is _____.

    b.   Let's use the value of each ratio to make multiplicative comparisons for each of the ratios we described here.

        i.   The number preferring new extra crunchy is _____ of the total number surveyed.

        ii.   The number preferring regular crunchy is _____ of the total number surveyed.

        iii.   The number preferring regular crunchy is _____ of those preferring new extra crunchy.

        iv.   The number preferring new extra crunchy is _____ of those preferring regular crunchy.

    c.   If the company is planning to produce 90,000 containers of crunchy peanut butter, how many of these containers should be the new extra crunchy variety, and how many of these containers should be the regular crunchy peanut butter? What would be helpful in solving this problem? Does one of our comparison statements above help us?

Try these next scenarios:

    d.   If the company decides to produce 2,000 containers of regular crunchy peanut butter, how many containers of new extra crunchy peanut butter would it produce?

    e.   If the company decides to produce 10,000 containers of new extra crunchy peanut butter, how many containers of regular crunchy peanut butter would it produce?

    f.   If the company decides to only produce 3,000 containers of new extra crunchy peanut butter, how many containers of regular crunchy peanut butter would it produce?

2015 Great Minds. eureka-math.org
G6-M1-SE-B1-1.3.1-01.2016

Lesson Summary

For a ratio $A:B$, we are often interested in the associated ratio $B:A$. Further, if A and B can both be measured in the same unit, we are often interested in the associated ratios $A:(A+B)$ and $B:(A+B)$.

For example, if Tom caught 3 fish and Kyle caught 5 fish, we can say:

The ratio of the number of fish Tom caught to the number of fish Kyle caught is $3:5$.

The ratio of the number of fish Kyle caught to the number of fish Tom caught is $5:3$.

The ratio of the number of fish Tom caught to the total number of fish the two boys caught is $3:8$.

The ratio of the number of fish Kyle caught to the total number of fish the two boys caught is $5:8$.

For the ratio $A:B$, where $B \neq 0$, the value of the ratio is the quotient $\dfrac{A}{B}$.

For example: For the ratio $6:8$, the value of the ratio is $\dfrac{6}{8}$ or $\dfrac{3}{4}$.

Problem Set

1. Maritza is baking cookies to bring to school and share with her friends on her birthday. The recipe requires 3 eggs for every 2 cups of sugar. To have enough cookies for all of her friends, Maritza determined she would need 12 eggs. If her mom bought 6 cups of sugar, does Maritza have enough sugar to make the cookies? Why or why not?

2. Hamza bought 8 gallons of brown paint to paint his kitchen and dining room. Unfortunately, when Hamza started painting, he thought the paint was too dark for his house, so he wanted to make it lighter. The store manager would not let Hamza return the paint but did inform him that if he used $\dfrac{1}{4}$ of a gallon of white paint mixed with 2 gallons of brown paint, he would get the shade of brown he desired. If Hamza decided to take this approach, how many gallons of white paint would Hamza have to buy to lighten the 8 gallons of brown paint?

This page intentionally left blank

Lesson 8: Equivalent Ratios Defined Through the Value of a Ratio

Classwork

Exercise 1

Circle any equivalent ratios from the list below.

      Ratio: $1:2$

      Ratio: $5:10$

      Ratio: $6:16$

      Ratio: $12:32$

Find the value of the following ratios, leaving your answer as a fraction, but rewrite the fraction using the largest possible unit.

      Ratio: $1:2$ Value of the Ratio:

      Ratio: $5:10$ Value of the Ratio:

      Ratio: $6:16$ Value of the Ratio:

      Ratio: $12:32$ Value of the Ratio:

What do you notice about the value of the equivalent ratios?

Exercise 2

Here is a theorem: If $A:B$ with $B \neq 0$ and $C:D$ with $D \neq 0$ are equivalent, then they have the same value: $\dfrac{A}{B} = \dfrac{C}{D}$.

This is essentially stating that if two ratios are equivalent, then their values are the same (when they have values).

Can you provide any counterexamples to the theorem above?

EUREKA
MATH™

©2015 Great Minds. eureka-math.org
G6-M1-SE-B1-1.3.1-01.2016

Exercise 3

Taivon is training for a duathlon, which is a race that consists of running and cycling. The cycling leg is longer than the running leg of the race, so while Taivon trains, he rides his bike more than he runs. During training, Taivon runs 4 miles for every 14 miles he rides his bike.

a.   Identify the ratio associated with this problem and find its value.

Use the value of each ratio to solve the following.

b.   When Taivon completed all of his training for the duathlon, the ratio of total number of miles he ran to total number of miles he cycled was $80:280$. Is this consistent with Taivon's training schedule? Explain why or why not.

c.   In one training session, Taivon ran 4 miles and cycled 7 miles. Did this training session represent an equivalent ratio of the distance he ran to the distance he cycled? Explain why or why not.

2015 Great Minds. eureka-math.org
G6-M1-SE-B1-1.3.1-01.2016

Lesson Summary

The *value of the ratio* $A:B$ is the quotient $\frac{A}{B}$ as long as B is not zero.

If two ratios are equivalent, then their values are the same (when they have values).

Problem Set

1. The ratio of the number of shaded sections to the number of unshaded sections is 4 to 2. What is the value of the ratio of the number of shaded pieces to the number of unshaded pieces?

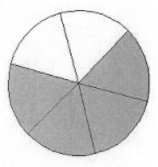

2. Use the value of the ratio to determine which ratios are equivalent to $7:15$.

    a. $21:45$

    b. $14:45$

    c. $3:5$

    d. $63:135$

3. Sean was at batting practice. He swung 25 times but only hit the ball 15 times.

    a. Describe and write more than one ratio related to this situation.

    b. For each ratio you created, use the value of the ratio to express one quantity as a fraction of the other quantity.

    c. Make up a word problem that a student can solve using one of the ratios and its value.

4. Your middle school has 900 students. $\frac{1}{3}$ of students bring their lunch instead of buying lunch at school. What is the value of the ratio of the number of students who do bring their lunch to the number of students who do not?

EUREKA MATH™

©2015 Great Minds. eureka-math.org
G6-M1-SE-B1-1.3.1-01.2016

This page intentionally left blank

Lesson 9: Tables of Equivalent Ratios

Classwork

Example 1

To make paper mache, the art teacher mixes water and flour. For every two cups of water, she needs to mix in three cups of flour to make the paste.

Find equivalent ratios for the ratio relationship 2 cups of water to 3 cups of flour. Represent the equivalent ratios in the table below:

Cups of Water	Cups of Flour

©2015 Great Minds. eureka-math.org
G6-M1-SE-B1-1.3.1-01.2016

Example 2

Javier has a new job designing websites. He is paid at a rate of $700 for every 3 pages of web content that he builds. Create a ratio table to show the total amount of money Javier has earned in ratio to the number of pages he has built.

Total Pages Built								
Total Money Earned								

Javier is saving up to purchase a used car that costs $4,200. How many web pages will Javier need to build before he can pay for the car?

Exercise 1

Spraying plants with cornmeal juice is a natural way to prevent fungal growth on the plants. It is made by soaking cornmeal in water, using two cups of cornmeal for every nine gallons of water. Complete the ratio table to answer the questions that follow.

Cups of Cornmeal	Gallons of Water

a. How many cups of cornmeal should be added to 45 gallons of water?

b. Paul has only 8 cups of cornmeal. How many gallons of water should he add if he wants to make as much cornmeal juice as he can?

c. What can you say about the values of the ratios in the table?

Exercise 2

James is setting up a fish tank. He is buying a breed of goldfish that typically grows to be 12 inches long. It is recommended that there be 1 gallon of water for every inch of fish length in the tank. What is the recommended ratio of gallons of water per fully-grown goldfish in the tank?

Complete the ratio table to help answer the following questions:

Number of Fish	Gallons of Water

a.   What size tank (in gallons) is needed for James to have 5 full-grown goldfish?

b.   How many fully-grown goldfish can go in a 40-gallon tank?

c.   What can you say about the values of the ratios in the table?

Lesson Summary

A ratio table is a table of pairs of numbers that form equivalent ratios.

Problem Set

Assume each of the following represents a table of equivalent ratios. Fill in the missing values. Then choose one of the tables and create a real-world context for the ratios shown in the table.

1.

	22
12	
16	44
	55
24	66

2.

	14
15	21
25	35
30	

3.

	34
	51
12	
15	85
18	102

©2015 Great Minds. eureka-math.org
G6-M1-SE-B1-1.3.1-01.2016

This page intentionally left blank

Lesson 10: The Structure of Ratio Tables—Additive and Multiplicative

Classwork

Exploratory Challenge

Imagine that you are making a fruit salad. For every quart of blueberries you add, you would like to put in 3 quarts of strawberries. Create three ratio tables that show the amounts of blueberries and strawberries you would use if you needed to make fruit salad for greater numbers of people.

Table 1 should contain amounts where you have added fewer than 10 quarts of blueberries to the salad.

Table 2 should contain amounts of blueberries between and including 10 and 50 quarts.

Table 3 should contain amounts of blueberries greater than or equal to 100 quarts.

Table 1	
Quarts of Blueberries	Quarts of Strawberries

Table 2	
Quarts of Blueberries	Quarts of Strawberries

Table 3	
Quarts of Blueberries	Quarts of Strawberries

©2015 Great Minds. eureka-math.org
G6-M1-SE-B1-1.3.1-01.2016

a.   Describe any patterns you see in the tables. Be specific in your descriptions.

b.   How are the amounts of blueberries and strawberries related to each other?

c.   How are the values in the Blueberries column related to each other?

d.   How are the values in the Strawberries column related to each other?

e.   If we know we want to add 7 quarts of blueberries to the fruit salad in Table 1, how can we use the table to help us determine how many strawberries to add?

2015 Great Minds. eureka-math.org
6-M1-SE-B1-1.3.1-01.2016

EUREKA
MATH™

f. If we know we used 70 quarts of blueberries to make our salad, how can we use a ratio table to find out how many quarts of strawberries were used?

Exercise 1

The following tables were made incorrectly. Find the mistakes that were made, create the correct ratio table, and state the ratio that was used to make the correct ratio table.

a.

Hours	Pay in Dollars
3	24
5	40
7	52
9	72

Hours	Pay in Dollars

Ratio _____

b.

Blue	Yellow
1	5
4	8
7	13
10	16

Blue	Yellow

Ratio _____

©2015 Great Minds. eureka-math.org
G6-M1-SE-B1-1.3.1-01.2016

Lesson Summary

Ratio tables are constructed in a special way.

Each pair of values in the table will be equivalent to the same ratio.

red	white
3	12
6	24
21	84

$6 : 24$ $21 : 84$

$1 : 4$ $1 : 4$

Repeated addition or multiplication can be used to create a ratio table.

The values in the first column can be multiplied by a constant value to get the values in the second column.

red		white
3	× 4	12
6	× 4	24
12	× 4	48
21	× 4	84

By just adding a certain number to the first entry of a ratio in the first column and adding the same number to the second entry in the second column, the new ratio formed is generally not equivalent to the original ratio. Instead, the numbers added to the entries must be related to the ratio used to make the table. However, if the entries in one column are multiplied by a certain number, multiplying the entries in the other column by the same number creates equivalent ratios.

red	white
3	12
6	24
12	48
21	84

x7   x7

2015 Great Minds. eureka-math.org
6-M1-SE-B1-1.3.1-01.2016

EUREKA
MATH

Problem Set

1.
   a.   Create a ratio table for making lemonade with a lemon juice-to-water ratio of $1:3$. Show how much lemon juice would be needed if you use 36 cups of water to make lemonade.

   b.   How is the value of the ratio used to create the table?

2.   Ryan made a table to show how much blue and red paint he mixed to get the shade of purple he will use to paint the room. He wants to use the table to make larger and smaller batches of purple paint.

Blue	Red
12	3
20	5
28	7
36	9

   a.   What ratio was used to create this table? Support your answer.

   b.   How are the values in each row related to each other?

   c.   How are the values in each column related to each other?

This page intentionally left blank

Lesson 11: Comparing Ratios Using Ratio Tables

Classwork

Example 1

Create four equivalent ratios (2 by scaling up and 2 by scaling down) using the ratio 30 to 80.

Write a ratio to describe the relationship shown in the table.

Hours	Number of Pizzas Sold
2	16
5	40
6	48
10	80

Exercise 1

The following tables show how many words each person can text in a given amount of time. Compare the rates of texting for each person using the ratio table.

Michaela

Minutes	3	5	7	9
Words	150	250	350	450

Jenna

Minutes	2	4	6	8
Words	90	180	270	360

Maria

Minutes	3	6	9	12
Words	120	240	360	480

Complete the table so that it shows Max has a texting rate of 55 words per minute.

Max

Minutes				
Words				

Exercise 2: Making Juice (Comparing Juice to Water)

a.   The tables below show the comparison of the amount of water to the amount of juice concentrate (JC) in grape juice made by three different people. Whose juice has the greatest water-to-juice concentrate ratio, and whose juice would taste strongest? Be sure to justify your answer.

Laredo's Juice			Franca's Juice			Milton's Juice		
Water	JC	Total	Water	JC	Total	Water	JC	Total
12	4	16	10	2	12	8	2	10
15	5	20	15	3	18	16	4	20
21	7	28	25	5	30	24	6	30
45	15	60	40	8	48	40	10	50

Put the juices in order from the juice containing the most water to the juice containing the least water.

_____   _____   _____

Explain how you used the values in the table to determine the order.

What ratio was used to create each table?

Laredo: _____   Franca: _____

Milton: _____

Explain how the ratio could help you compare the juices.

b.   The next day, each of the three people made juice again, but this time they were making apple juice. Whose juice has the greatest water-to-juice concentrate ratio, and whose juice would taste the strongest? Be sure to justify your answer.

Laredo's Juice		
Water	JC	Total
12	2	14
18	3	21
30	5	35
42	7	49

Franca's Juice		
Water	JC	Total
15	6	21
20	8	28
35	14	49
50	20	70

Milton's Juice		
Water	JC	Total
16	6	22
24	9	33
40	15	55
64	24	88

Put the juices in order from the strongest apple taste to the weakest apple taste.

_____   _____   _____

Explain how you used the values in the table to determine the order.

What ratio was used to create each table?

Laredo: _____   Franca: _____

Milton: _____

©2015 Great Minds. eureka-math.org
G6-M1-SE-B1-1.3.1-01.2016

Explain how the ratio could help you compare the juices.

How was this problem different than the grape juice questions in part (a)?

c.   Max and Sheila are making orange juice. Max has mixed 15 cups of water with 4 cups of juice concentrate. Sheila has made her juice by mixing 8 cups water with 3 cups of juice concentrate. Compare the ratios of juice concentrate to water using ratio tables. State which beverage has a higher juice concentrate-to-water ratio.

d.   Victor is making recipes for smoothies. His first recipe calls for 2 cups of strawberries and 7 cups of other ingredients. His second recipe says that 3 cups of strawberries are combined with 9 cups of other ingredients. Which smoothie recipe has more strawberries compared to other ingredients? Use ratio tables to justify your answer.

2015 Great Minds. eureka-math.org
6-M1-SE-B1-1.3.1-01.2016

EUREKA
MATH™

Lesson Summary

Ratio tables can be used to compare two ratios.

Look for equal amounts in a row or column to compare the second amount associated with it.

3	6	12	30
7	14	28	70

10	25	30	45
16	40	48	72

The values of the tables can also be extended in order to get comparable amounts. Another method would be to compare the values of the ratios by writing the values of the ratios as fractions and then using knowledge of fractions to compare the ratios.

When ratios are given in words, creating a table of equivalent ratios helps in comparing the ratios.

12: 35 compared to 8: 20

Quantity 1	12	24	36	48
Quantity 2	35	70	105	140

Quantity 1	8	56
Quantity 2	20	140

Problem Set

1. Sarah and Eva were swimming.

   a. Use the ratio tables below to determine who the faster swimmer is.

   Sarah

Time (min)	3	5	12	17
Distance (meters)	75	125	300	425

   Eva

Time (min)	2	7	10	20
Distance (meters)	52	182	260	520

   b. Explain the method that you used to determine your answer.

2. A 120 lb. person would weigh about 20 lb. on the earth's moon. A 150 lb. person would weigh 28 lb. on Io, a moon of Jupiter. Use ratio tables to determine which moon would make a person weigh the most.

This page intentionally left blank

Lesson 12: From Ratio Tables to Double Number Line Diagrams

Classwork

Exercise 2

The amount of sugary beverages Americans consume is a leading health concern. For a given brand of cola, a 12 oz. serving of cola contains about 40 g of sugar. Complete the ratio table, using the given ratio to find equivalent ratios.

Cola (ounces)		12	
Sugar (grams)		40	

Exercise 3

A 1 L bottle of cola contains approximately 34 fluid ounces. How many grams of sugar would be in a 1 L bottle of the cola? Explain and show how to arrive at the solution.

Exercise 4

A school cafeteria has a restriction on the amount of sugary drinks available to students. Drinks may not have more than 25 g of sugar. Based on this restriction, what is the largest size cola (in ounces) the cafeteria can offer to students?

©2015 Great Minds. eureka-math.org
G6-M1-SE-B1-1.3.1-01.2016

Exercise 5

Shontelle solves three math problems in four minutes.

a. Use this information to complete the table below.

Number of Questions	3	6	9	12	15	18	21	24	27	30
Number of Minutes										

b. Shontelle has soccer practice on Thursday evening. She has a half hour before practice to work on her math homework and to talk to her friends. She has 20 math skill-work questions for homework, and she wants to complete them before talking with her friends. How many minutes will Shontelle have left after completing her math homework to talk to her friends?

Use a double number line diagram to support your answer, and show all work.

EUREKA MATH™

2015 Great Minds. eureka-math.org
G6-M1-SE-B1-1.3.1-01.2016

Lesson Summary

A *double number line* is a representation of a ratio relationship using a pair of parallel number lines. One number line is drawn above the other so that the zeros of each number line are aligned directly with each other. Each ratio in a ratio relationship is represented on the double number line by always plotting the first entry of the ratio on one of the number lines and plotting the second entry on the other number line so that the second entry is aligned with the first entry.

Problem Set

1. While shopping, Kyla found a dress that she would like to purchase, but it costs $52.25 more than she has. Kyla charges $5.50 an hour for babysitting. She wants to figure out how many hours she must babysit to earn $52.25 to buy the dress. Use a double number line to support your answer.

2. Frank has been driving at a constant speed for 3 hours, during which time he traveled 195 miles. Frank would like to know how long it will take him to complete the remaining 455 miles, assuming he maintains the same constant speed. Help Frank determine how long the remainder of the trip will take. Include a table or diagram to support your answer.

©2015 Great Minds. eureka-math.org
G6-M1-SE-B1-1.3.1-01.2016

This page intentionally left blank

Double Number Line Reproducible

©2015 Great Minds. eureka-math.org
G6-M1-SE-B1-1.3.1-01.2016

This page intentionally left blank

Lesson 13: From Ratio Tables to Equations Using the Value of a Ratio

Classwork

Exercise 1

Jorge is mixing a special shade of orange paint. He mixed 1 gallon of red paint with 3 gallons of yellow paint.

Based on this ratio, which of the following statements are true?

- $\frac{3}{4}$ of a 4-gallon mix would be yellow paint.

- Every 1 gallon of yellow paint requires $\frac{1}{3}$ gallon of red paint.

- Every 1 gallon of red paint requires 3 gallons of yellow paint.

- There is 1 gallon of red paint in a 4-gallon mix of orange paint.

- There are 2 gallons of yellow paint in an 8-gallon mix of orange paint.

Use the space below to determine if each statement is true or false.

Exercise 2

Based on the information on red and yellow paint given in Exercise 1, complete the table below.

Red Paint (R)	Yellow Paint (Y)	Relationship
	3	$3 = 1 \times 3$
2		
	9	$9 = 3 \times 3$
	12	
5		

Exercise 3

a. Jorge now plans to mix red paint and blue paint to create purple paint. The color of purple he has decided to make combines red paint and blue paint in the ratio $4:1$. If Jorge can only purchase paint in one gallon containers, construct a ratio table for all possible combinations for red and blue paint that will give Jorge no more than 25 gallons of purple paint.

Blue (B)	Red (R)	Relationship

Write an equation that will let Jorge calculate the amount of red paint he will need for any given amount of blue paint.

2015 Great Minds. eureka-math.org
6-M1-SE-B1-1.3.1-01.2016

EUREKA
MATH™

Write an equation that will let Jorge calculate the amount of blue paint he will need for any given amount of red paint.

If Jorge has 24 gallons of red paint, how much blue paint will he have to use to create the desired color of purple?

If Jorge has 24 gallons of blue paint, how much red paint will he have to use to create the desired color of purple?

b. Using the same relationship of red to blue from above, create a table that models the relationship of the three colors blue, red, and purple (total) paint. Let B represent the number of gallons of blue paint, let R represent the number of gallons of red paint, and let T represent the total number of gallons of (purple) paint. Then write an equation that models the relationship between the blue paint and the total paint, and answer the questions.

Blue (B)	Red (R)	Total Paint (T)

Equation:

Value of the ratio of total paint to blue paint:

How is the value of the ratio related to the equation?

Exercise 4

During a particular U.S. Air Force training exercise, the ratio of the number of men to the number of women was 6∶1. Use the ratio table provided below to create at least two equations that model the relationship between the number of men and the number of women participating in this training exercise.

Women (W)	Men (M)

Equations:

If 200 women participated in the training exercise, use one of your equations to calculate the number of men who participated.

Exercise 5

Malia is on a road trip. During the first five minutes of Malia's trip, she sees 18 cars and 6 trucks. Assuming this ratio of cars to trucks remains constant over the duration of the trip, complete the ratio table using this comparison. Let T represent the number of trucks she sees, and let C represent the number of cars she sees.

Trucks (T)	Cars (C)
1	
3	
	18
12	
	60

What is the value of the ratio of the number of cars to the number of trucks?

What equation would model the relationship between cars and trucks?

At the end of the trip, Malia had counted 1,254 trucks. How many cars did she see?

2015 Great Minds. eureka-math.org
6-M1-SE-B1-1.3.1-01.2016

Exercise 6

Kevin is training to run a half-marathon. His training program recommends that he run for 5 minutes and walk for 1 minute. Let R represent the number of minutes running, and let W represent the number of minutes walking.

Minutes Running (R)		10	20		50
Minutes Walking (W)	1	2		8	

What is the value of the ratio of the number of minutes walking to the number of minutes running?

What equation could you use to calculate the minutes spent walking if you know the minutes spent running?

Lesson Summary

The value of a ratio can be determined using a ratio table. This value can be used to write an equation that also represents the ratio.

Example:

1	4
2	8
3	12
4	16

The multiplication table can be a valuable resource to use in seeing ratios. Different rows can be used to find equivalent ratios.

Problem Set

A cookie recipe calls for 1 cup of white sugar and 3 cups of brown sugar.

Make a table showing the comparison of the amount of white sugar to the amount of brown sugar.

White Sugar (W)	Brown Sugar (B)

1. Write the value of the ratio of the amount of white sugar to the amount of brown sugar.

2. Write an equation that shows the relationship of the amount of white sugar to the amount of brown sugar.

3. Explain how the value of the ratio can be seen in the table.

4. Explain how the value of the ratio can be seen in the equation.

2015 Great Minds. eureka-math.org
6-M1-SE-B1-1.3.1-01.2016

Using the same recipe, compare the amount of white sugar to the amount of total sugars used in the recipe.

Make a table showing the comparison of the amount of white sugar to the amount of total sugar.

White Sugar (W)	Total Sugar (T)

5.   Write the value of the ratio of the amount of total sugar to the amount of white sugar.

6.   Write an equation that shows the relationship of total sugar to white sugar.

This page intentionally left blank

Lesson 14: From Ratio Tables, Equations, and Double Number Line Diagrams to Plots on the Coordinate Plane

Classwork

Kelli is traveling by train with her soccer team from Yonkers, NY to Morgantown, WV for a tournament. The distance between Yonkers and Morgantown is 400 miles. The total trip will take 8 hours. The train schedule is provided below:

Leaving Yonkers, New York	
Destination	**Distance**
Allentown, PA	100 miles
Carlisle, PA	200 miles
Berkeley Springs, WV	300 miles
Morgantown, WV	400 miles

Leaving Morgantown, WV	
Destination	**Distance**
Berkeley Springs, WV	100 miles
Carlisle, PA	200 miles
Allentown, PA	300 miles
Yonkers, NY	400 miles

Exercises

1.   Create a table to show the time it will take Kelli and her team to travel from Yonkers to each town listed in the schedule assuming that the ratio of the amount of time traveled to the distance traveled is the same for each city. Then, extend the table to include the cumulative time it will take to reach each destination on the ride home.

Hours	Miles

2. Create a double number line diagram to show the time it will take Kelli and her team to travel from Yonkers to each town listed in the schedule. Then, extend the double number line diagram to include the cumulative time it will take to reach each destination on the ride home. Represent the ratio of the distance traveled on the round trip to the amount of time taken with an equation.

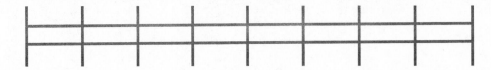

Using the information from the double number line diagram, how many miles would be traveled in one hour?

_____

How do you know?

_____

_____

Example 1

Dinner service starts once the train is 250 miles away from Yonkers. What is the minimum time the players will have to wait before they can have their meal?

Hours	Miles	Ordered Pairs

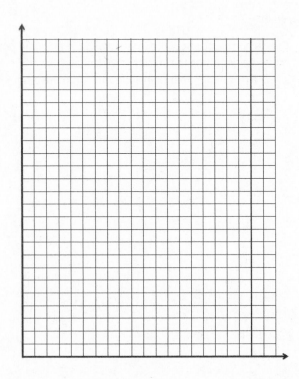

EUREKA MATH

Lesson Summary

A ratio table, equation, or double number line diagram can be used to create ordered pairs. These ordered pairs can then be graphed on a coordinate plane as a representation of the ratio.

Example:

Equation: $y = 3x$

x	y
0	0
1	3
2	6
3	9

Ordered Pairs

(x, y)
$(0, 0)$
$(1, 3)$
$(2, 6)$
$(3, 9)$

Y = 3x

Problem Set

1. Complete the table of values to find the following:

    Find the number of cups of sugar needed if for each pie Karrie makes, she has to use 3 cups of sugar.

Pies	Cups of Sugar
1	
2	
3	
4	
5	
6	

Use a graph to represent the relationship.

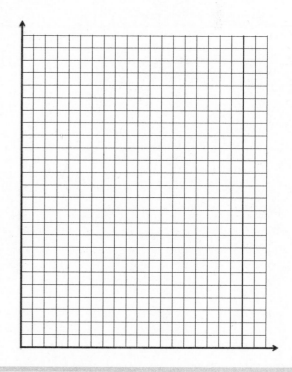

EUREKA MATH

Lesson 14:   From Ratio Tables, Equations, and Double Number Line Diagrams to
            Plots on the Coordinate Plane

S.71

©2015 Great Minds. eureka-math.org
G6-M1-SE-B1-1.3.1-01.2016

Create a double number line diagram to show the relationship.

2.   Write a story context that would be represented by the ratio $1:4$.

Complete a table of values for this equation and graph.

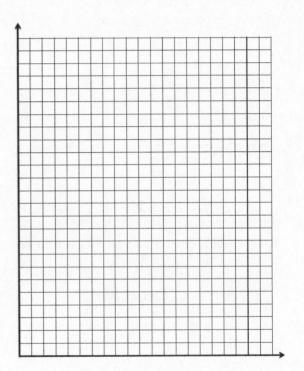

Lesson 14:   From Ratio Tables, Equations, and Double Number Line Diagrams to
Plots on the Coordinate Plane

2015 Great Minds. eureka-math.org
6-M1-SE-B1-1.3.1-01.2016

**EUREKA
MATH**

Lesson 15: A Synthesis of Representations of Equivalent Ratio Collections

Classwork

Exploratory Challenge

At the end of this morning's news segment, the local television station highlighted area pets that need to be adopted. The station posted a specific website on the screen for viewers to find more information on the pets shown and the adoption process. The station producer checked the website two hours after the end of the broadcast and saw that the website had 24 views. One hour after that, the website had 36 views.

Exercise 1

Create a table to determine how many views the website probably had one hour after the end of the broadcast based on how many views it had two and three hours after the end of the broadcast. Using this relationship, predict how many views the website will have 4, 5, and 6 hours after the end of the broadcast.

Exercise 2

What is the constant number, c, that makes these ratios equivalent?

Using an equation, represent the relationship between the number of views, v, the website received and the number of hours, h, after this morning's news broadcast.

Exercise 3

Use the table created in Exercise 1 to identify sets of ordered pairs that can be graphed.

Exercise 4

Use the ordered pairs you created to depict the relationship between hours and number of views on a coordinate plane. Label your axes and create a title for the graph. Do the points you plotted lie on a line?

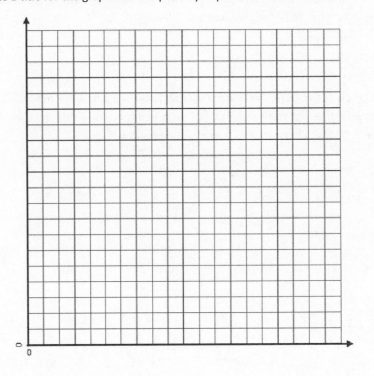

EUREKA
MATH™

2015 Great Minds. eureka-math.org
6-M1-SE-B1-1.3.1-01.2016

Exercise 5

Predict how many views the website will have after twelve hours. Use at least two representations (e.g., tape diagram, table, double number line diagram) to justify your answer.

Exercise 6

Also on the news broadcast, a chef from a local Italian restaurant demonstrated how he makes fresh pasta daily for his restaurant. The recipe for his pasta is below:

3 eggs, beaten

1 teaspoon salt

2 cups all-purpose flour

2 tablespoons water

2 tablespoons vegetable oil

Determine the ratio of the number of tablespoons of water to the number of eggs.

Provided the information in the table below, complete the table to determine ordered pairs. Use the ordered pairs to graph the relationship of the number of tablespoons of water to the number of eggs.

Tablespoons of Water	Number of Eggs
2	
4	
6	
8	
10	
12	

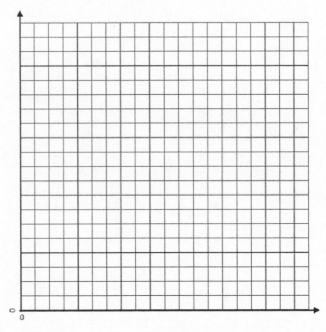

What would you have to do to the graph in order to find how many eggs would be needed if the recipe was larger and called for 16 tablespoons of water?

Demonstrate on your graph.

How many eggs would be needed if the recipe called for 16 tablespoons of water?

Lesson 15:   A Synthesis of Representations of Equivalent Ratio Collections

2015 Great Minds. eureka-math.org
G6-M1-SE-B1-1.3.1-01.2016

EUREKA MATH™

Exercise 7

Determine how many tablespoons of water will be needed if the chef is making a large batch of pasta and the recipe increases to 36 eggs. Support your reasoning using at least one diagram you find applies best to the situation, and explain why that tool is the best to use.

©2015 Great Minds. eureka-math.org
G6-M1-SE-B1-1.3.1-01.2016

Lesson Summary

There are several ways to represent the same collection of equivalent ratios. These include ratio tables, tape diagrams, double number line diagrams, equations, and graphs on coordinate planes.

Problem Set

1. The producer of the news station posted an article about the high school's football championship ceremony on a new website. The website had 500 views after four hours. Create a table to show how many views the website would have had after the first, second, and third hours after posting, if the website receives views at the same rate. How many views would the website receive after 5 hours?

2. Write an equation that represents the relationship from Problem 1. Do you see any connections between the equations you wrote and the ratio of the number of views to the number of hours?

3. Use the table in Problem 1 to make a list of ordered pairs that you could plot on a coordinate plane.

4. Graph the ordered pairs on a coordinate plane. Label your axes and create a title for the graph.

5. Use multiple tools to predict how many views the website would have after 12 hours.

2015 Great Minds. eureka-math.org
6-M1-SE-B1-1.3.1-01.2016

Graph Reproducible

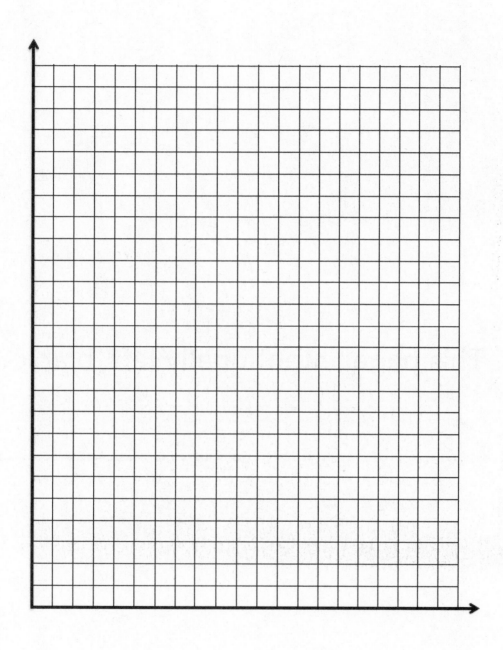

©2015 Great Minds. eureka-math.org
G6-M1-SE-B1-1.3.1-01.2016

This page intentionally left blank

Lesson 16: From Ratios to Rates

Classwork

Ratios can be transformed to rates and unit rates.

Example: Introduction to Rates and Unit Rates

Diet cola was on sale last week; it cost $10 for every 4 packs of diet cola.

    a.   How much do 2 packs of diet cola cost?

    b.   How much does 1 pack of diet cola cost?

Exploratory Challenge

    a.   Teagan went to Gamer Realm to buy new video games. Gamer Realm was having a sale: $65 for 4 video games. He bought 3 games for himself and one game for his friend, Diego, but Teagan does not know how much Diego owes him for the one game. What is the unit price of the video games? What is the rate unit?

©2015 Great Minds. eureka-math.org
G6-M1-SE-B1-1.3.1-01.2016

b.   Four football fans took turns driving the distance from New York to Oklahoma to see a big game. Each driver set the cruise control during his or her portion of the trip, enabling him or her to travel at a constant speed. The group changed drivers each time they stopped for gas and recorded their driving times and distances in the table below.

Fan	Distance (miles)	Time (hours)
Andre	208	4
Matteo	456	8
Janaye	300	6
Greyson	265	5

Use the given data to answer the following questions.

i.   What two quantities are being compared?

ii.   What is the ratio of the two quantities for Andre's portion of the trip? What is the associated rate?

Andre's Ratio: _____   Andre's Rate: _____

iii.   Answer the same two questions in part (ii) for the other three drivers.

Matteo's Ratio: _____   Matteo's Rate: _____

Janaye's Ratio: _____   Janaye's Rate: _____

Greyson's Ratio: _____   Greyson's Rate: _____

iv.   For each driver in parts (ii) and (iii), circle the unit rate and put a box around the rate unit.

EUREKA
MATH™

2015 Great Minds. eureka-math.org
6-M1-SE-B1-1.3.1-01.2016

c. A publishing company is looking for new employees to type novels that will soon be published. The publishing company wants to find someone who can type at least 45 words per minute. Dominique discovered she can type at a constant rate of 704 words in 16 minutes. Does Dominique type at a fast enough rate to qualify for the job? Explain why or why not.

Lesson Summary

A *rate* is a quantity that describes a ratio relationship between two types of quantities.

For example, 15 miles/hour is a rate that describes a ratio relationship between hours and miles: If an object is traveling at a constant 15 miles/hour, then after 1 hour it has gone 15 miles, after 2 hours it has gone 30 miles, after 3 hours it has gone 45 miles, and so on.

When a rate is written as a measurement, the *unit rate* is the measure (i.e., the numerical part of the measurement). For example, when the rate of speed of an object is written as the measurement 15 miles/hour, the number 15 is the unit rate. The *unit of measurement* is miles/hour, which is read as "miles per hour."

Problem Set

The Scott family is trying to save as much money as possible. One way to cut back on the money they spend is by finding deals while grocery shopping; however, the Scott family needs help determining which stores have the better deals.

1. At Grocery Mart, strawberries cost $2.99 for 2 lb., and at Baldwin Hills Market strawberries are $3.99 for 3 lb.

    a. What is the unit price of strawberries at each grocery store? If necessary, round to the nearest penny.

    b. If the Scott family wanted to save money, where should they go to buy strawberries? Why?

2. Potatoes are on sale at both Grocery Mart and Baldwin Hills Market. At Grocery Mart, a 5 lb. bag of potatoes cost $2.85, and at Baldwin Hills Market a 7 lb. bag of potatoes costs $4.20. Which store offers the best deal on potatoes? How do you know? How much better is the deal?

2015 Great Minds. eureka-math.org
6-M1-SE-B1-1.3.1-01.2016

Lesson 17: From Rates to Ratios

Classwork

Given a rate, you can calculate the unit rate and associated ratios. Recognize that all ratios associated with a given rate are equivalent because they have the same value.

Example 1

Write each ratio as a rate.

a. The ratio of miles to the number of hours is 434 to 7.

b. The ratio of the number of laps to the number of minutes is 5 to 4.

Example 2

a. Complete the model below using the ratio from Example 1, part (b).

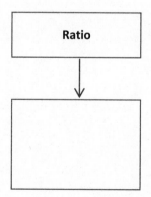

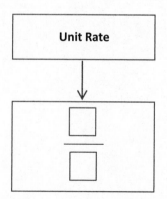

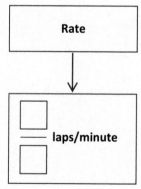

b.   Complete the model below now using the rate listed below.

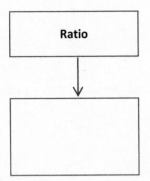   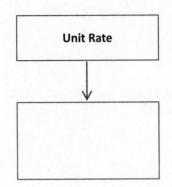   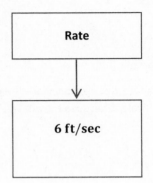

Examples 3–6

3.   Dave can clean pools at a constant rate of $\frac{3}{5}$ pools/hour.

   a.   What is the ratio of the number of pools to the number of hours?

   b.   How many pools can Dave clean in 10 hours?

   c.   How long does it take Dave to clean 15 pools?

EUREKA
MATH™

2015 Great Minds. eureka-math.org
G6-M1-SE-B1-1.3.1-01.2016

4. Emeline can type at a constant rate of $\frac{1}{4}$ pages/minute.

    a. What is the ratio of the number of pages to the number of minutes?

    b. Emeline has to type a 5-page article but only has 18 minutes until she reaches the deadline. Does Emeline have enough time to type the article? Why or why not?

    c. Emeline has to type a 7-page article. How much time will it take her?

5. Xavier can swim at a constant speed of $\frac{5}{3}$ meters/second.

    a. What is the ratio of the number of meters to the number of seconds?

    b. Xavier is trying to qualify for the National Swim Meet. To qualify, he must complete a 100-meter race in 55 seconds. Will Xavier be able to qualify? Why or why not?

    c. Xavier is also attempting to qualify for the same meet in the 200-meter event. To qualify, Xavier would have to complete the race in 130 seconds. Will Xavier be able to qualify in this race? Why or why not?

©2015 Great Minds. eureka-math.org
G6-M1-SE-B1-1.3.1-01.2016

6. The corner store sells apples at a rate of 1.25 dollars per apple.

   a. What is the ratio of the amount in dollars to the number of apples?

   b. Akia is only able to spend $10 on apples. How many apples can she buy?

   c. Christian has $6 in his wallet and wants to spend it on apples. How many apples can Christian buy?

2015 Great Minds. eureka-math.org
©6-M1-SE-B1-1.3.1-01.2016

EUREKA
MATH™

Lesson Summary

A rate of $\frac{2}{3}$ gal/min corresponds to the unit rate of $\frac{2}{3}$ and also corresponds to the ratio $2:3$.

All ratios associated with a given rate are equivalent because they have the same value.

Problem Set

1. Once a commercial plane reaches the desired altitude, the pilot often travels at a cruising speed. On average, the cruising speed is 570 miles/hour. If a plane travels at this cruising speed for 7 hours, how far does the plane travel while cruising at this speed?

2. Denver, Colorado often experiences snowstorms resulting in multiple inches of accumulated snow. During the last snow storm, the snow accumulated at $\frac{4}{5}$ inch/hour. If the snow continues at this rate for 10 hours, how much snow will accumulate?

This page intentionally left blank

Lesson 18: Finding a Rate by Dividing Two Quantities

Classwork

Mathematical Modeling Exercises

1. At Fun Burger, the Burger Master can make hamburgers at a rate of 4 burgers/minute. In order to address the heavy volume of customers, he needs to continue at this rate for 30 minutes. If he continues to make hamburgers at this pace, how many hamburgers will the Burger Master make in 30 minutes?

2. Chandra is an editor at the New York Gazette. Her job is to read each article before it is printed in the newspaper. If Chandra can read 10 words/second, how many words can she read in 60 seconds?

Exercises

Use the table below to write down your work and answers for the stations.

1.
2.
3.
4.
5.
6.

EUREKA
MATH™

2015 Great Minds. eureka-math.org
6-M1-SE-B1-1.3.1-01.2016

Lesson Summary

We can convert measurement units using rates. The information can be used to further interpret the problem. Here is an example:

$$\left(5\,\frac{\text{gal}}{\text{min}}\right) \cdot (10\text{ min}) = \frac{5\text{ gal}}{1\,\cancel{\text{min}}} \cdot 10\,\cancel{\text{min}} = 50\text{ gal}$$

Problem Set

1. Enguun earns $17 per hour tutoring student-athletes at Brooklyn University.

    a. If Enguun tutored for 12 hours this month, how much money did she earn this month?

    b. If Enguun tutored for 19.5 hours last month, how much money did she earn last month?

2. The Piney Creek Swim Club is preparing for the opening day of the summer season. The pool holds 22,410 gallons of water, and water is being pumped in at 540 gallons per hour. The swim club has its first practice in 42 hours. Will the pool be full in time? Explain your answer.

This page intentionally left blank

Lesson 19: Comparison Shopping—Unit Price and Related Measurement Conversions

Classwork

Analyze tables, graphs, and equations in order to compare rates.

Examples: Creating Tables from Equations

1. The ratio of cups of blue paint to cups of red paint is $1:2$, which means for every cup of blue paint, there are two cups of red paint. In this case, the equation would be red $= 2 \times$ blue, or $r = 2b$, where b represents the amount of blue paint and r represents the amount of red paint. Make a table of values.

2. Ms. Siple is a librarian who really enjoys reading. She can read $\frac{3}{4}$ of a book in one day. This relationship can be represented by the equation days$= \frac{3}{4}$ books, which can be written as $d = \frac{3}{4}b$, where b represents the number of books and d represents the number of days.

©2015 Great Minds. eureka-math.org
G6-M1-SE-B1-1.3.1-01.2016

Exercises

1. Bryan and ShaNiece are both training for a bike race and want to compare who rides his or her bike at a faster rate. Both bikers use apps on their phones to record the time and distance of their bike rides. Bryan's app keeps track of his route on a table, and ShaNiece's app presents the information on a graph. The information is shown below.

Bryan:

Number of Hours	0	3	6
Number of Miles	0	75	150

ShaNiece:

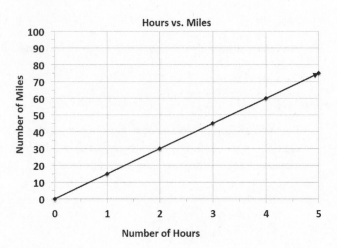

a. At what rate does each biker travel? Explain how you arrived at your answer.

b. ShaNiece wants to win the bike race. Make a new graph to show the speed ShaNiece would have to ride her bike in order to beat Bryan.

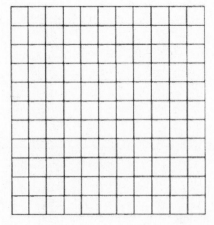

S.96   Lesson 19:   Comparison Shopping—Unit Price and Related Measurement
                               Conversions

2015 Great Minds. eureka-math.org
G6-M1-SE-B1-1.3.1-01.2016

EUREKA
MATH

2. Braylen and Tyce both work at a department store and are paid by the hour. The manager told the boys they both earn the same amount of money per hour, but Braylen and Tyce did not agree. They each kept track of how much money they earned in order to determine if the manager was correct. Their data is shown below.

Braylen: $m = 10.50h$ where h represents the number of hours worked and m represents the amount of money Braylen was paid.

Tyce:

Number of Hours	0	3	6
Money in Dollars	0	34.50	69

a. How much did each person earn in one hour?

b. Was the manager correct? Why or why not?

3. Claire and Kate are entering a cup stacking contest. Both girls have the same strategy: stack the cups at a constant rate so that they do not slow down at the end of the race. While practicing, they keep track of their progress, which is shown below.

Claire:

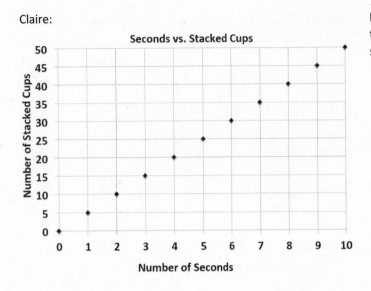

Kate: $c = 4t$, where t represents the amount of time in seconds and c represents the number of stacked cups.

a. At what rate does each girl stack her cups during the practice sessions?

b. Kate notices that she is not stacking her cups fast enough. What would Kate's equation look like if she wanted to stack cups faster than Claire?

2015 Great Minds. eureka-math.org
G6-M1-SE-B1-1.3.1-01.2016

Lesson Summary

When comparing rates and ratios, it is best to find the unit rate.

Comparing unit rates can happen across tables, graphs, and equations.

Problem Set

Victor was having a hard time deciding which new vehicle he should buy. He decided to make the final decision based on the gas efficiency of each car. A car that is more gas efficient gets more miles per gallon of gas. When he asked the manager at each car dealership for the gas mileage data, he received two different representations, which are shown below.

Vehicle 1: Legend

Gallons of Gas	4	8	12
Number of Miles	72	144	216

Vehicle 2: Supreme

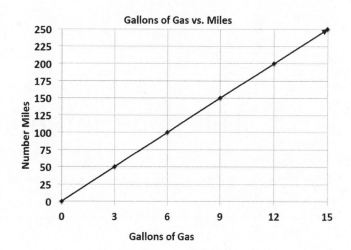

1. If Victor based his decision only on gas efficiency, which car should he buy? Provide support for your answer.

2. After comparing the Legend and the Supreme, Victor saw an advertisement for a third vehicle, the Lunar. The manager said that the Lunar can travel about 289 miles on a tank of gas. If the gas tank can hold 17 gallons of gas, is the Lunar Victor's best option? Why or why not?

EUREKA MATH

Lesson 19:   Comparison Shopping—Unit Price and Related Measurement
             Conversions

S.99

©2015 Great Minds. eureka-math.org
G6-M1-SE-B1-1.3.1-01.2016

This page intentionally left blank

Lesson 20: Comparison Shopping—Unit Price and Related Measurement Conversions

Classwork

An activity will be completed in order to gain confidence in comparing rates on tables, graphs, and equations.

Example 1: Notes from Exit Ticket

Take notes from the discussion in the space provided below.

Notes:

Exploratory Challenge

a. Mallory is on a budget and wants to determine which cereal is a better buy. A 10-ounce box of cereal costs $2.79, and a 13-ounce box of the same cereal costs $3.99.

   i.   Which box of cereal should Mallory buy?

   ii. What is the difference between the two unit prices?

b. Vivian wants to buy a watermelon. Kingston's Market has 10-pound watermelons for $3.90, but the Farmer's Market has 12-pound watermelons for $4.44.

   i. Which market has the best price for watermelon?

   ii. What is the difference between the two unit prices?

c. Mitch needs to purchase soft drinks for a staff party. He is trying to figure out if it is cheaper to buy the 12-pack of soda or the 20-pack of soda. The 12-pack of soda costs $3.99, and the 20-pack of soda costs $5.48.

   i. Which pack should Mitch choose?

   ii. What is the difference in cost between single cans of soda from each of the two packs?

d. Mr. Steiner needs to purchase 60 AA batteries. A nearby store sells a 20-pack of AA batteries for $12.49 and a 12-pack of the same batteries for $7.20.

   i. Would it be less expensive for Mr. Steiner to purchase the batteries in 20-packs or 12-packs?

   ii. What is the difference between the costs of one battery from each pack?

e. The table below shows the amount of calories Mike burns as he runs.

Number of Miles Ran	3	6	9	12
Number of Calories Burned	360	720		1,440

Fill in the missing part of the table.

f. Emilio wants to buy a new motorcycle. He wants to compare the gas efficiency for each motorcycle before he makes a purchase. The dealerships presented the data below.

Sports Motorcycle:

Number of Gallons of Gas	5	10	15	20
Number of Miles	287.5	575	862.5	1,150

Leisure Motorcycle:

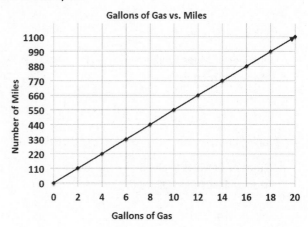

Which motorcycle is more gas efficient and by how much?

g. Milton Middle School is planning to purchase a new copy machine. The principal has narrowed the choice to two models: SuperFast Deluxe and Quick Copies. He plans to purchase the machine that copies at the fastest rate. Use the information below to determine which copier the principal should choose.

SuperFast Deluxe:

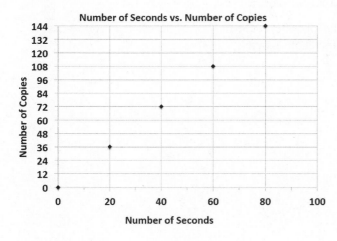

Quick Copies:

$$c = 1.5t$$

(where t represents the amount of time in seconds, and c represents the number of copies)

h. Elijah and Sean are participating in a walk-a-thon. Each student wants to calculate how much money he would make from his sponsors at different points of the walk-a-thon. Use the information in the tables below to determine which student would earn more money if they both walked the same distance. How much more money would that student earn per mile?

Elijah's Sponsor Plan:

Number of Miles Walked	7	14	21	28
Money Earned in Dollars	35	70	105	140

Sean's Sponsor Plan:

Number of Miles Walked	6	12	18	24
Money Earned in Dollars	33	66	99	132

i. Gerson is going to buy a new computer to use for his new job and also to download movies. He has to decide between two different computers. How many more kilobytes does the faster computer download in one second?

Choice 1: The rate of download is represented by the following equation: $k = 153t$, where t represents the amount of time in seconds, and k represents the number of kilobytes.

Choice 2: The rate of download is represented by the following equation: $k = 150t$, where t represents the amount of time in seconds, and k represents the number of kilobytes.

Lesson 20:   Comparison Shopping—Unit Price and Related Measurement
Conversions

2015 Great Minds. eureka-math.org
G6-M1-SE-B1-1.3.1-01.2016

j. Zyearaye is trying to decide which security system company he will make more money working for. Use the graphs below that show Zyearaye's potential commission rate to determine which company will pay Zyearaye more commission. How much more commission would Zyearaye earn by choosing the company with the better rate?

Superior Security:

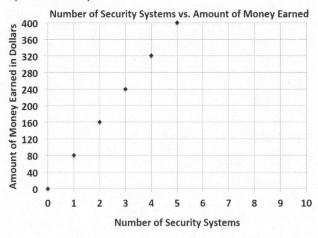

Top Notch Security:

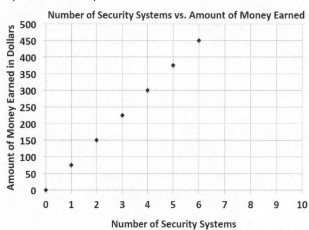

k. Emilia and Miranda are sisters, and their mother just signed them up for a new cell phone plan because they send too many text messages. Using the information below, determine which sister sends the most text messages. How many more text messages does this sister send per week?

Emilia:

Number of Weeks	3	6	9	12
Number of Text Messages	1,200	2,400	3,600	4,800

Miranda: $m = 410w$, where w represents the number of weeks, and m represents the number of text messages.

©2015 Great Minds. eureka-math.org
G6-M1-SE-B1-1.3.1-01.2016

Lesson Summary

Unit Rate can be located in tables, graphs, and equations.

- Table—the unit rate is the value of the first quantity when the second quantity is **1**.

- Graphs—the unit rate is the value of r at the point $(1, r)$.

- Equation—the unit rate is the constant number in the equation. For example, the unit rate in $r = 3b$ is **3**.

Problem Set

The table below shows the amount of money Gabe earns working at a coffee shop.

Number of Hours Worked	3	6	9	12
Money Earned (in dollars)	40.50	81.00	121.50	162.00

1. How much does Gabe earn per hour?

2. Jordan is another employee at the same coffee shop. He has worked there longer than Gabe and earns $3 more per hour than Gabe. Complete the table below to show how much Jordan earns.

Number of Hours Worked	4	8	12	16
Money Earned (in dollars)				

3. Serena is the manager of the coffee shop. The amount of money she earns is represented by the equation $m = 21h$, where h is the number of hours Serena works, and m is the amount of money she earns. How much more money does Serena make an hour than Gabe? Explain your thinking.

4. Last month, Jordan received a promotion and became a manager. He now earns the same amount as Serena. How much more money does he earn per hour now that he is a manager than he did before his promotion? Explain your thinking.

2015 Great Minds. eureka-math.org
G6-M1-SE-B1-1.3.1-01.2016

EUREKA
MATH

Lesson 21: Getting the Job Done—Speed, Work, and Measurement Units

Classwork

Conversion tables contain ratios that can be used to convert units of length, weight, or capacity. You must multiply the given number by the ratio that compares the two units.

Opening Exercise

Identify the ratios that are associated with conversions between feet, inches, and yards.

12 inches = _____ foot; the ratio of inches to feet is _____.

1 foot = _____ inches; the ratio of feet to inches is _____.

3 feet = _____ yard; the ratio of feet to yards is _____.

1 yard = _____ feet; the ratio of yards to feet is _____.

Example 1

Work with your partner to find out how many feet are in 48 inches. Make a ratio table that compares feet and inches. Use the conversion rate of 12 inches per foot or $\frac{1}{12}$ foot per inch.

Example 2

How many grams are in 6 kilograms? Again, make a record of your work before using the calculator. The rate would be 1,000 grams per kilogram. The unit rate would be 1,000.

Exercise 1

How many cups are in 5 quarts? As always, make a record of your work before using the calculator. The rate would be 4 cups per quart. The unit rate would be 4.

Exercise 2

How many quarts are in 10 cups?

2015 Great Minds. eureka-math.org
6-M1-SE-B1-1.3.1-01.2016

Lesson Summary

Conversion tables contain ratios that can be used to convert units of length, weight, or capacity. You must multiply the given number by the ratio that compares the two units.

Problem Set

1. 7 ft. = _____ in.

2. 100 yd. = _____ ft.

3. 25 m = _____ cm

4. 5 km = _____ m

5. 96 oz. = _____ lb.

6. 2 mi. = _____ ft.

7. 2 mi. = _____ yd.

8. 32 fl. oz. = _____ c.

9. 1,500 mL = _____ L

10. 6 g = _____ mg

11. Beau buys a 3-pound bag of trail mix for a hike. He wants to make one-ounce bags for his friends with whom he is hiking. How many one-ounce bags can he make?_____

12. The maximum weight for a truck on the New York State Thruway is 40 tons. How many pounds is this?_____

13. Claudia's skis are 150 centimeters long. How many meters is this?_____

14. Claudia's skis are 150 centimeters long. How many millimeters is this?_____

15. Write your own problem, and solve it. Be ready to share the question tomorrow.

This page intentionally left blank

U.S. Customary Length	Conversion
Inch (in.)	$1 \text{ in.} = \dfrac{1}{12} \text{ ft.}$
Foot (ft.)	$1 \text{ ft.} = 12 \text{ in.}$
Yard (yd.)	$1 \text{ yd.} = 3 \text{ ft.}$ $1 \text{ yd.} = 36 \text{ in.}$
Mile (mi.)	$1 \text{ mi.} = 1{,}760 \text{ yd.}$ $1 \text{ mi.} = 5{,}280 \text{ ft.}$

Metric Length	Conversion
Centimeter (cm)	$1 \text{ cm} = 10 \text{ mm}$
Meter (m)	$1 \text{ m} = 100 \text{ cm}$ $1 \text{ m} = 1{,}000 \text{ mm}$
Kilometer (km)	$1 \text{ km} = 1{,}000 \text{ m}$

U.S. Customary Weight	Conversion
Pound (lb.)	$1 \text{ lb.} = 16 \text{ oz.}$
Ton (T.)	$1 \text{ T.} = 2{,}000 \text{ lb.}$

Metric Capacity	Conversion
Liter (L)	$1 \text{ L} = 1{,}000 \text{ ml}$
Kiloliter (kL)	$1 \text{ kL} = 1{,}000 \text{ L}$

U.S. Customary Capacity	Conversion
Cup (c.)	$1 \text{ c.} = 8 \text{ fluid ounces}$
Pint (pt.)	$1 \text{ pt.} = 2 \text{ c.}$
Quart (qt.)	$1 \text{ qt.} = 4 \text{ c.}$ $1 \text{ qt.} = 2 \text{ pt.}$ $1 \text{ qt.} = 32 \text{ fluid ounces}$
Gallon (gal.)	$1 \text{ gal.} = 4 \text{ qt.}$ $1 \text{ gal.} = 8 \text{ pt.}$ $1 \text{ gal.} = 16 \text{ c.}$ $1 \text{ gal.} = 128 \text{ fluid ounces}$

Metric Mass	Conversion
Gram (g)	$1 \text{ g} = 1{,}000 \text{ mg}$
Kilogram (kg)	$1 \text{ kg} = 1{,}000 \text{ g}$

©2015 Great Minds. eureka-math.org
G6-M1-SE-B1-1.3.1-01.2016

This page intentionally left blank

Lesson 22: Getting the Job Done—Speed, Work, and Measurement Units

Classwork

If an object is moving at a constant rate of speed for a certain amount of time, it is possible to find how far the object went by multiplying the rate and the time. In mathematical language, we say, distance = rate · time.

Example 1

Walker: Substitute the walker's distance and time into the equation and solve for the rate of speed.

distance = rate · time

$d = r \cdot t$

Hint: Consider the units that you want to end up with. If you want to end up with the rate (feet/second), then divide the distance (feet) by time (seconds).

Runner: Substitute the runner's time and distance into the equation to find the rate of speed.

distance = rate · time

$d = r \cdot t$

Example 2

<u>Part 1</u>: Chris Johnson ran the 40-yard dash in 4.24 seconds. What is the rate of speed? Round any answer to the nearest hundredth.

distance $=$ rate $\cdot$ time

$d = r \cdot t$

<u>Part 2</u>: In Lesson 21, we converted units of measure using unit rates. If the runner were able to run at a constant rate, how many yards would he run in an hour? This problem can be solved by breaking it down into two steps. Work with a partner, and make a record of your calculations.

a.   How many yards would he run in one minute?

b.   How many yards would he run in one hour?

We completed that problem in two separate steps, but it is possible to complete this same problem in one step. We can multiply the yards per second by the seconds per minute, then by the minutes per hour.

$$\underline{\hspace{1.5cm}} \frac{\text{yards}}{\text{second}} \cdot 60 \, \frac{\text{seconds}}{\text{minute}} \cdot 60 \, \frac{\text{minutes}}{\text{hour}} = \underline{\hspace{2.5cm}} \text{ yards in one hour}$$

Cross out any units that are in both the numerator and denominator in the expression because these cancel each other out.

<u>Part 3</u>: How many miles did the runner travel in that hour? Round your response to the nearest tenth.

Cross out any units that are in both the numerator and denominator in the expression because they cancel out.

2015 Great Minds. eureka-math.org
G6-M1-SE-B1-1.3.1-01.2016

Exercises: Road Trip

Exercise 1

I drove my car on cruise control at 65 miles per hour for 3 hours without stopping. How far did I go?

$d = r \cdot t$

$d =$ _____ $\dfrac{\text{miles}}{\text{hour}} \cdot$ _____ hours

Cross out any units that are in both the numerator and denominator in the expression because they cancel out.

$d =$ _____ miles

Exercise 2

On the road trip, the speed limit changed to 50 miles per hour in a construction zone. Traffic moved along at a constant rate (50 mph), and it took me 15 minutes (0.25 hours) to get through the zone. What was the distance of the construction zone? (Round your response to the nearest hundredth of a mile.)

$d = r \cdot t$

$d =$ _____ $\dfrac{\text{miles}}{\text{hour}} \cdot$ _____ hours

EUREKA MATH™

©2015 Great Minds. eureka-math.org
G6-M1-SE-B1-1.3.1-01.2016

Lesson Summary

Distance, rate, and time are related by the formula $d = r \cdot t$.

Knowing any two of the values allows the calculation of the third.

Problem Set

1. If Adam's plane traveled at a constant speed of 375 miles per hour for six hours, how far did the plane travel?

2. A Salt March Harvest Mouse ran a 360 centimeter straight course race in 9 seconds. How fast did it run?

3. Another Salt Marsh Harvest Mouse took 7 seconds to run a 350 centimeter race. How fast did it run?

4. A slow boat to China travels at a constant speed of 17.25 miles per hour for 200 hours. How far was the voyage?

5. The Sopwith Camel was a British, First World War, single-seat, biplane fighter introduced on the Western Front in 1917. Traveling at its top speed of 110 mph in one direction for $2\frac{1}{2}$ hours, how far did the plane travel?

6. A world-class marathon runner can finish 26.2 miles in 2 hours. What is the rate of speed for the runner?

7. Banana slugs can move at 17 cm per minute. If a banana slug travels for 5 hours, how far will it travel?

2015 Great Minds. eureka-math.org
G6-M1-SE-B1-1.3.1-01.2016

Lesson 23: Problem Solving Using Rates, Unit Rates, and Conversions

Classwork

- If work is being done at a constant rate by one person, and at a different constant rate by another person, both rates can be converted to their unit rates and then compared directly.
- "Work" can include jobs done in a certain time period, rates of running or swimming, etc.

Example 1: Fresh-Cut Grass

Suppose that on a Saturday morning you can cut 3 lawns in 5 hours, and your friend can cut 5 lawns in 8 hours. Who is cutting lawns at a faster rate?

$$\frac{3 \text{ lawns}}{5 \text{ hours}} = \frac{\rule{2cm}{0.4pt} \text{ lawns}}{1 \quad \text{hour}}$$

$$\frac{5 \text{ lawns}}{8 \text{ hours}} = \frac{\rule{2cm}{0.4pt} \text{ lawns}}{1 \quad \text{hour}}$$

Example 2: Restaurant Advertising

$$\frac{\rule{2cm}{0.4pt} \text{ menus}}{\rule{2cm}{0.4pt} \text{ hours}} = \frac{\rule{2cm}{0.4pt} \text{ menus}}{1 \quad \text{hour}}$$

$$\frac{\rule{2cm}{0.4pt} \text{ menus}}{\rule{2cm}{0.4pt} \text{ hours}} = \frac{\rule{2cm}{0.4pt} \text{ menus}}{1 \quad \text{hour}}$$

EUREKA MATH™

©2015 Great Minds. eureka-math.org
G6-M1-SE-B1-1.3.1-01.2016

Example 3: Survival of the Fittest

$$\frac{\underline{\hspace{1cm}}\ \text{feet}}{\underline{\hspace{1cm}}\ \text{seconds}} = \frac{\underline{\hspace{1cm}}\ \text{feet}}{1\ \text{second}} \qquad\qquad \frac{\underline{\hspace{1cm}}\ \text{feet}}{\underline{\hspace{1cm}}\ \text{seconds}} = \frac{\underline{\hspace{1cm}}\ \text{feet}}{1\ \text{second}}$$

Example 4: Flying Fingers

$$\underline{\hspace{3cm}} = \underline{\hspace{3cm}} \qquad\qquad \underline{\hspace{3cm}} = \underline{\hspace{3cm}}$$

2015 Great Minds. eureka-math.org
6-M1-SE-B1-1.3.1-01.2016

EUREKA
MATH

> **Lesson Summary**
>
> - Rate problems, including constant rate problems, always count or measure something happening per unit of time. The time is always in the denominator.
> - Sometimes the units of time in the denominators of the rates being compared are not the same. One must be converted to the other before calculating the unit rate of each.

Problem Set

1. Who walks at a faster rate: someone who walks 60 feet in 10 seconds or someone who walks 42 feet in 6 seconds?

2. Who walks at a faster rate: someone who walks 60 feet in 10 seconds or someone who takes 5 seconds to walk 25 feet? Review the lesson summary before answering.

3. Which parachute has a slower decent: a red parachute that falls 10 feet in 4 seconds or a blue parachute that falls 12 feet in 6 seconds?

4. During the winter of 2012–2013, Buffalo, New York received 22 inches of snow in 12 hours. Oswego, New York received 31 inches of snow over a 15-hour period. Which city had a heavier snowfall rate? Round your answers to the nearest hundredth.

5. A striped marlin can swim at a rate of 70 miles per hour. Is this a faster or slower rate than a sailfish, which takes 30 minutes to swim 40 miles?

6. One math student, John, can solve 6 math problems in 20 minutes while another student, Juaquine, can solve the same 6 math problems at a rate of 1 problem per 4 minutes. Who works faster?

©2015 Great Minds. eureka-math.org
G6-M1-SE-B1-1.3.1-01.2016

This page intentionally left blank

Lesson 24: Percent and Rates per 100

Classwork

Exercise 1

Robb's Fruit Farm consists of 100 acres on which three different types of apples grow. On 25 acres, the farm grows Empire apples. McIntosh apples grow on 30% of the farm. The remainder of the farm grows Fuji apples. Shade in the grid below to represent the portion of the farm each type of apple occupies. Use a different color for each type of apple. Create a key to identify which color represents each type of apple.

Color Key

Empire _____

McIntosh _____

Fuji _____

Part-to-Whole Ratio

_____

_____

_____

Exercise 2

The shaded portion of the grid below represents the portion of a granola bar remaining.

What percent does each block of granola bar represent?

What percent of the granola bar remains?

What other ways can we represent this percent?

0.01	0.01	0.01	0.01	0.01	0.01	0.01	0.01	0.01	0.01
0.01	0.01	0.01	0.01	0.01	0.01	0.01	0.01	0.01	0.01
0.01	0.01	0.01	0.01	0.01	0.01	0.01	0.01	0.01	0.01
0.01	0.01	0.01	0.01	0.01	0.01	0.01	0.01	0.01	0.01
0.01	0.01	0.01	0.01	0.01	0.01	0.01	0.01	0.01	0.01
0.01	0.01	0.01	0.01	0.01	0.01	0.01	0.01	0.01	0.01
0.01	0.01	0.01	0.01	0.01	0.01	0.01	0.01	0.01	0.01
0.01	0.01	0.01	0.01	0.01	0.01	0.01	0.01	0.01	0.01
0.01	0.01	0.01	0.01	0.01	0.01	0.01	0.01	0.01	0.01
0.01	0.01	0.01	0.01	0.01	0.01	0.01	0.01	0.01	0.01

©2015 Great Minds. eureka-math.org
G6-M1-SE-B1-1.3.1-01.2016

Exercise 3

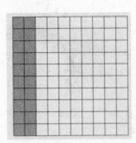
a.

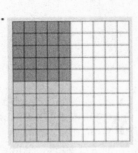
b.

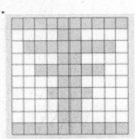
c.

a. For each figure shown, represent the gray shaded region as a percent of the whole figure. Write your answer as a decimal, fraction, and percent.

Picture (a)	Picture (b)	Picture (c)

b. What ratio is being modeled in each picture?

c. How are the ratios and percentages related?

EUREKA
MATH™

2015 Great Minds. eureka-math.org
G6-M1-SE-B1-1.3.1-01.2016

Exercise 4

Each relationship below compares the shaded portion (the part) to the entire figure (the whole). Complete the table.

Percentage	Decimal	Fraction	Ratio	Model
6%			6: 100	
60%				
600%				
32%				

©2015 Great Minds. eureka-math.org
G6-M1-SE-B1-1.3.1-01.2016

	0.55		
	$\dfrac{9}{10}$		

Exercise 5

Mr. Brown shares with the class that 70% of the students got an A on the English vocabulary quiz. If Mr. Brown has 100 students, create a model to show how many of the students received an A on the quiz.

What fraction of the students received an A on the quiz?

How could we represent this amount using a decimal?

How are the decimal, the fraction, and the percent all related?

EUREKA
MATH™

2015 Great Minds. eureka-math.org
G6-M1-SE-B1-1.3.1-01.2016

Exercise 6

Marty owns a lawn mowing service. His company, which consists of three employees, has 100 lawns to mow this week. Use the 10 × 10 grid to model how the work could have been distributed between the three employees.

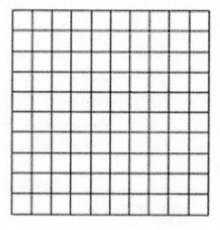

Worker	Percentage	Fraction	Decimal
Employee 1			
Employee 2			
Employee 3			

Color over the name with the same color you used in the diagram.

©2015 Great Minds. eureka-math.org
G6-M1-SE-B1-1.3.1-01.2016

Lesson Summary

One percent is the number $\frac{1}{100}$ and is written as 1%.

Percentages can be used as rates. For example, 30% of a quantity means $\frac{30}{100}$ times the quantity.

We can create models of percents. One example would be to shade a 10×10 grid. Each square in a 10×10 grid represents 1% or 0.01.

Problem Set

1. Marissa just bought 100 acres of land. She wants to grow apple, peach, and cherry trees on her land. Color the model to show how the acres could be distributed for each type of tree. Using your model, complete the table.

Tree	Percentage	Fraction	Decimal
Apple			
Peach			
Cherry			

EUREKA
MATH™

2015 Great Minds. eureka-math.org
6-M1-SE-B1-1.3.1-01.2016

2. After renovations on Kim's bedroom, only 30 percent of one wall is left without any décor. Shade the grid below to represent the space that is left to decorate.

   a. What does each block represent?

   b. What percent of this wall has been decorated?

0.01	0.01	0.01	0.01	0.01	0.01	0.01	0.01	0.01	0.01
0.01	0.01	0.01	0.01	0.01	0.01	0.01	0.01	0.01	0.01
0.01	0.01	0.01	0.01	0.01	0.01	0.01	0.01	0.01	0.01
0.01	0.01	0.01	0.01	0.01	0.01	0.01	0.01	0.01	0.01
0.01	0.01	0.01	0.01	0.01	0.01	0.01	0.01	0.01	0.01
0.01	0.01	0.01	0.01	0.01	0.01	0.01	0.01	0.01	0.01
0.01	0.01	0.01	0.01	0.01	0.01	0.01	0.01	0.01	0.01
0.01	0.01	0.01	0.01	0.01	0.01	0.01	0.01	0.01	0.01
0.01	0.01	0.01	0.01	0.01	0.01	0.01	0.01	0.01	0.01
0.01	0.01	0.01	0.01	0.01	0.01	0.01	0.01	0.01	0.01

©2015 Great Minds. eureka-math.org
G6-M1-SE-B1-1.3.1-01.2016

This page intentionally left blank

Lesson 25: A Fraction as a Percent

Classwork

Example 1

Sam says 50% of the vehicles are cars. Give three different reasons or models that prove or disprove Sam's statement.
Models can include tape diagrams, 10×10 grids, double number lines, etc.

How is the fraction of cars related to the percent?

Use a model to prove that the fraction and percent are equivalent.

What other fractions or decimals also represent 60%?

Example 2

A survey was taken that asked participants whether or not they were happy with their job. An overall score was given. 300 of the participants were unhappy while 700 of the participants were happy with their job. Give a part-to-whole fraction for comparing happy participants to the whole. Then write a part-to-whole fraction of the unhappy participants to the whole. What percent were happy with their job, and what percent were unhappy with their job?

Happy _____ _____   Unhappy _____ _____
          Fraction   Percent   Fraction   Percent

Create a model to justify your answer.

Exercise 1

Renita claims that a score of 80% means that she answered $\frac{4}{5}$ of the problems correctly. She drew the following picture to support her claim:

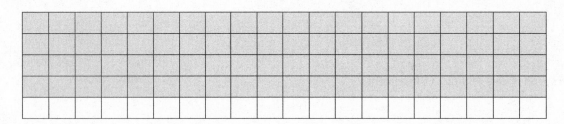

Is Renita correct? _____ Why or why not?

How could you change Renita's picture to make it easier for Renita to see why she is correct or incorrect?

2015 Great Minds. eureka-math.org
G6-M1-SE-B1-1.3.1-01.2016

Exercise 2

Use the diagram to answer the following questions.

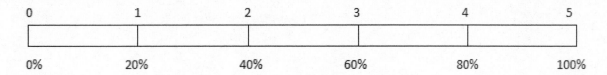

80% is what fraction of the whole quantity?

$\dfrac{1}{5}$ is what percent of the whole quantity?

50% is what fraction of the whole quantity?

1 is what percent of the whole quantity?

EUREKA
MATH™

©2015 Great Minds. eureka-math.org
G6-M1-SE-B1-1.3.1-01.2016

Exercise 3

Maria completed $\frac{3}{4}$ of her workday. Create a model that represents what percent of the workday Maria has worked.

What percent of her workday does she have left?

How does your model prove that your answer is correct?

Exercise 4

Matthew completed $\frac{5}{8}$ of his workday. What decimal would also describe the portion of the workday he has finished?

How can you use the decimal to get the percent of the workday Matthew has completed?

2015 Great Minds. eureka-math.org
6-M1-SE-B1-1.3.1-01.2016

Exercise 5

Complete the conversions from fraction to decimal to percent.

Fraction	Decimal	Percent
$\dfrac{1}{8}$		
	0.35	
		84.5%
	0.325	
$\dfrac{2}{25}$		

Exercise 6

Choose one of the rows from the conversion table in Exercise 5, and use models to prove your answers. (Models could include a 10 × 10 grid, a tape diagram, a double number line, etc.)

©2015 Great Minds. eureka-math.org
G6-M1-SE-B1-1.3.1-01.2016

Lesson Summary

Fractions, decimals, and percentages are all related.

To change a fraction to a percentage, you can scale up or scale down so that 100 is in the denominator.

Example:

$$\frac{9}{20} = \frac{9 \times 5}{20 \times 5} = \frac{45}{100} = 45\%$$

There may be times when it is more beneficial to convert a fraction to a percent by first writing the fraction in decimal form.

Example:

$$\frac{5}{8} = 0.625 = 62.5 \text{ hundredths} = 62.5\%$$

Models, like tape diagrams and number lines, can also be used to model the relationships.

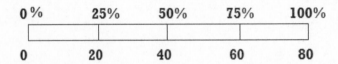

The diagram shows that $\frac{20}{80} = 25\%$.

Problem Set

1. Use the 10 × 10 grid to express the fraction $\frac{11}{20}$ as a percent.

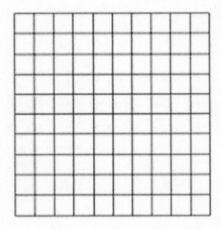

2. Use a tape diagram to relate the fraction $\frac{11}{20}$ to a percent.

3. How are the diagrams related?

4. What decimal is also related to the fraction?

5. Which diagram is the most helpful for converting the fraction to a decimal? _____ Explain why.

EUREKA
MATH

2015 Great Minds. eureka-math.org
G6-M1-SE-B1-1.3.1-01.2016

10 × 10 Grid Reproducible

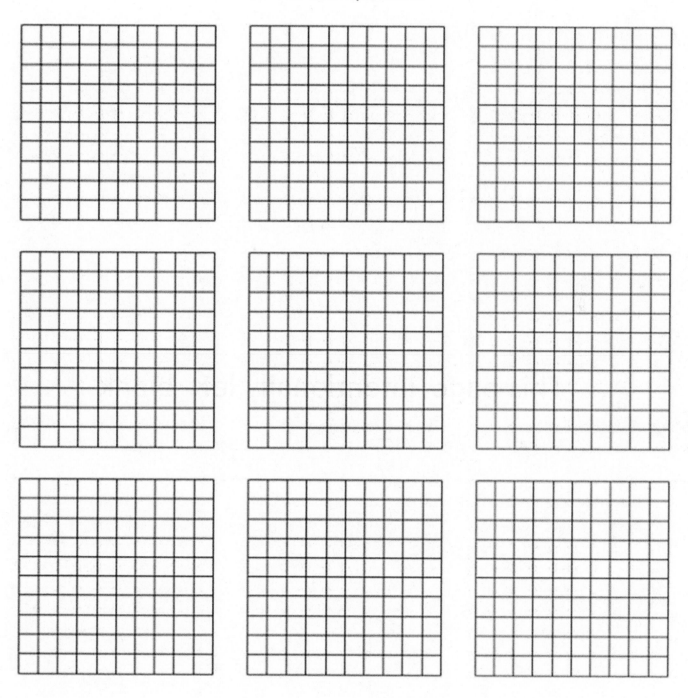

©2015 Great Minds. eureka-math.org
G6-M1-SE-B1-1.3.1-01.2016

This page intentionally left blank

Lesson 26: Percent of a Quantity

Classwork

Example 1

Five of the 25 girls on Alden Middle School's soccer team are seventh-grade students. Find the percentage of seventh graders on the team. Show two different ways of solving for the answer. One of the methods must include a diagram or picture model.

Example 2

Of the 25 girls on the Alden Middle School soccer team, 40% also play on a travel team. How many of the girls on the middle school team also play on a travel team?

Example 3

The Alden Middle School girls' soccer team won 80% of its games this season. If the team won 12 games, how many games did it play? Solve the problem using at least two different methods.

2015 Great Minds. eureka-math.org
G6-M1-SE-B1-1.3.1-01.2016

Exercises

1. There are 60 animal exhibits at the local zoo. What percent of the zoo's exhibits does each animal class represent?

Exhibits by Animal Class	Number of Exhibits	Percent of the Total Number of Exhibits
Mammals	30	
Reptiles & Amphibians	15	
Fish & Insects	12	
Birds	3	

2. A sweater is regularly $32. It is 25% off the original price this week.

    a. Would the amount the shopper saved be considered the part, whole, or percent?

    b. How much would a shopper save by buying the sweater this week? Show two methods for finding your answer.

3. A pair of jeans was 30% off the original price. The sale resulted in a $24 discount.

    a. Is the original price of the jeans considered the whole, part, or percent?

    b. What was the original cost of the jeans before the sale? Show two methods for finding your answer.

4. Purchasing a TV that is 20% off will save $180.

    a. Name the different parts with the words: PART, WHOLE, PERCENT.

    _____   _____   _____

          20% off $180   Original Price

    b. What was the original price of the TV? Show two methods for finding your answer.

2015 Great Minds. eureka-math.org
G6-M1-SE-B1-1.3.1-01.2016

EUREKA
MATH

Lesson Summary

Models and diagrams can be used to solve percent problems. Tape diagrams, 10 × 10 grids, double number line diagrams, and others can be used in a similar way to using them with ratios to find the percent, the part, or the whole.

Problem Set

1. What is 15% of 60? Create a model to prove your answer.

2. If 40% of a number is 56, what was the original number?

3. In a 10 × 10 grid that represents 800, one square represents _____.
    Use the grids below to represent 17% and 83% of 800.

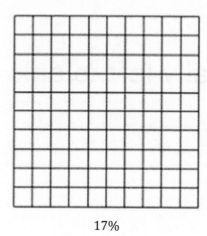

17%

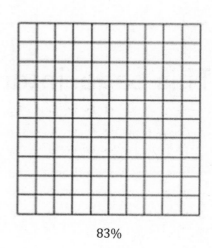

83%

17% of 800 is _____.   83% of 800 is _____.

This page intentionally left blank

Lesson 27: Solving Percent Problems

Classwork

Example 1

Solve the following three problems.

Write the words PERCENT, WHOLE, or PART under each problem to show which piece you were solving for.

60% of 300 = _____   60% of _____ = 300   60 out of 300 = _____%

_____   _____   _____

How did your solving method differ with each problem?

EUREKA MATH™

©2015 Great Minds. eureka-math.org
G6-M1-SE-B1-1.3.1-01.2016

Exercise

Use models, such as 10×10 grids, ratio tables, tape diagrams, or double number line diagrams, to solve the following situation.

Priya is doing her back-to-school shopping. Calculate all of the missing values in the table below, rounding to the nearest penny, and calculate the total amount Priya will spend on her outfit after she receives the indicated discounts.

	Shirt (25% discount)	Pants (30% discount)	Shoes (15% discount)	Necklace (10% discount)	Sweater (20% discount)
Original Price	$44			$20	
Amount of Discount		$15	$9		$7

What is the total cost of Priya's outfit?

Lesson Summary

Percent problems include the part, whole, and percent. When one of these values is missing, we can use tables, diagrams, and models to solve for the missing number.

Problem Set

1. Mr. Yoshi has 75 papers. He graded 60 papers, and he had a student teacher grade the rest. What percent of the papers did each person grade?

2. Mrs. Bennett has graded 20% of her 150 students' papers. How many papers does she still need to finish grading?

This page intentionally left blank

Lesson 28: Solving Percent Problems

Classwork

Example

If an item is discounted 20%, the sale price is what percent of the original price?

If the original price of the item is $400, what is the dollar amount of the discount?

How much is the sale price?

©2015 Great Minds. eureka-math.org
G6-M1-SE-B1-1.3.1-01.2016

Exercise

The following items were bought on sale. Complete the missing information in the table.

Item	Original Price	Sale Price	Amount of Discount	Percent Saved	Percent Paid
Television		$800		20%	
Sneakers	$80			25%	
Video Games		$54			90%
MP3 Player		$51.60		40%	
Book			$2.80		80%
Snack Bar		$1.70	$0.30		

2015 Great Minds. eureka-math.org
6-M1-SE-B1-1.3.1-01.2016

Lesson Summary

Percent problems include the part, whole, and percent. When one of these values is missing, we can use tables, diagrams, and models to solve for the missing number.

Problem Set

1. The Sparkling House Cleaning Company has cleaned 28 houses this week. If this number represents 40% of the total number of houses the company is contracted to clean, how many total houses will the company clean by the end of the week?

2. Joshua delivered 30 hives to the local fruit farm. If the farmer has paid to use 5% of the total number of Joshua's hives, how many hives does Joshua have in all?

This page intentionally left blank

Lesson 29: Solving Percent Problems

Classwork

Exploratory Challenge 1

Claim: To find 10% of a number, all you need to do is move the decimal to the left once.

Use at least one model to solve each problem (e.g., tape diagram, table, double number line diagram, 10×10 grid).

a. Make a prediction. Do you think the claim is true or false? _____ Explain why.

b. Determine 10% of 300. _____

c. Find 10% of 80. _____

d. Determine 10% of 64. _____

e. Find 10% of 5. _____

f. 10% of _____ is 48.

g. 10% of _____ is 6.

h.   Gary read 34 pages of a 340-page book. What percent did he read?

i.   Micah read 16 pages of his book. If this is 10% of the book, how many pages are in the book?

j.   Using the solutions to the problems above, what conclusions can you make about the claim?

Exploratory Challenge 2

Claim: If an item is already on sale, and then there is another discount taken off the new price, this is the same as taking the sum of the two discounts off the original price.

Use at least one model to solve each problem (e.g., tape diagram, table, double number line diagram, 10×10 grid).

a.   Make a prediction. Do you think the claim is true or false?_____ Explain.

b.   Sam purchased 3 games for $140 after a discount of 30%. What was the original price?

2015 Great Minds. eureka-math.org
G6-M1-SE-B1-1.3.1-01.2016

c. If Sam had used a 20% off coupon and opened a frequent shopper discount membership to save 10%, would the games still have a total of $140?

d. Do you agree with the claim? _____ Explain why or why not. Create a new example to help support your claim.

Lesson Summary

Percent problems have three parts: whole, part, percent.

Percent problems can be solved using models such as ratio tables, tape diagrams, double number line diagrams, and 10×10 grids.

Problem Set

1. Henry has 15 lawns mowed out of a total of 60 lawns. What percent of the lawns does Henry still have to mow?

2. Marissa got an 85% on her math quiz. She had 34 questions correct. How many questions were on the quiz?

3. Lucas read 30% of his book containing 480 pages. What page is he going to read next?

2015 Great Minds. eureka-math.org
G6-M1-SE-B1-1.3.1-01.2016

EUREKA
MATH™

Student Edition

Eureka Math
Grade 6
Module 2

Special thanks go to the Gordon A. Cain Center and to the Department of Mathematics at Louisiana State University for their support in the development of *Eureka Math*.

For a free *Eureka Math* Teacher
Resource Pack, Parent Tip
Sheets, and more please
visit www.Eureka.tools

Published by the non-profit Great Minds

Copyright © 2015 Great Minds. No part of this work may be reproduced, sold, or commercialized, in whole or in part, without written permission from Great Minds. Non-commercial use is licensed pursuant to a Creative Commons Attribution-NonCommercial-ShareAlike 4.0 license; for more information, go to http://greatminds.net/maps/math/copyright. "Great Minds" and "Eureka Math" are registered trademarks of Great Minds.

Printed in the U.S.A.
This book may be purchased from the publisher at eureka-math.org
10 9 8 7 6 5 4 3 2

ISBN 978-1-63255-312-6

Lesson 1: Interpreting Division of a Fraction by a Whole Number—Visual Models

Classwork

Opening Exercise

A

Write a division sentence to solve each problem.

1. 8 gallons of batter are poured equally into 4 bowls. How many gallons of batter are in each bowl?

2. 1 gallon of batter is poured equally into 4 bowls. How many gallons of batter are in each bowl?

Write a division sentence *and* draw a model to solve.

3. 3 gallons of batter are poured equally into 4 bowls. How many gallons of batter are in each bowl?

B

Write a multiplication sentence to solve each problem.

1. One fourth of an 8-gallon pail is poured out. How many gallons are poured out?

2. One fourth of a 1-gallon pail is poured out. How many gallons are poured out?

Write a multiplication sentence *and* draw a model to solve.

3. One fourth of a 3-gallon pail is poured out. How many gallons are poured out?

Example 1

$\frac{3}{4}$ gallon of batter is poured equally into 2 bowls. How many gallons of batter are in each bowl?

Example 2

$\frac{3}{4}$ pan of lasagna is shared equally by 6 friends. What fraction of the pan will each friend get?

Example 3

A rope of length $\frac{2}{5}$ m is cut into 4 equal cords. What is the length of each cord?

©2015 Great Minds. eureka-math.org
G6-M2-SE-B1-1.3.1-01.2016

Exercises 1–6

Fill in the blanks to complete the equation. Then, find the quotient and draw a model to support your solution.

1. $\dfrac{1}{2} \div 3 = \dfrac{\square}{2} \times \dfrac{1}{2}$

2. $\dfrac{1}{3} \div 4 = \dfrac{1}{4} \times \dfrac{1}{\square}$

Find the value of each of the following.

3. $\dfrac{1}{4} \div 5$

4. $\dfrac{3}{5} \div 5$

5. $\dfrac{1}{5} \div 4$

EUREKA MATH™

©2015 Great Minds. eureka-math.org
G6-M2-SE-B1-1.3.1-01.2016

Solve. Draw a model to support your solution.

6. $\dfrac{3}{5}$ pt. of juice is poured equally into 6 glasses. How much juice is in each glass?

EUREKA
MATH™

©2015 Great Minds. eureka-math.org
G6-M2-SE-B1-1.3.1-01.2016

Problem Set

Find the value of each of the following in its simplest form.

1.

    a. $\dfrac{1}{3} \div 4$ b. $\dfrac{2}{5} \div 4$ c. $\dfrac{4}{7} \div 4$

2.

    a. $\dfrac{2}{5} \div 3$ b. $\dfrac{5}{6} \div 5$ c. $\dfrac{5}{8} \div 10$

3.

    a. $\dfrac{6}{7} \div 3$ b. $\dfrac{10}{8} \div 5$ c. $\dfrac{20}{6} \div 2$

4. 4 loads of stone weigh $\dfrac{2}{3}$ ton. Find the weight of 1 load of stone.

5. What is the width of a rectangle with an area of $\dfrac{5}{8}$ in² and a length of 10 inches?

6. Lenox ironed $\dfrac{1}{4}$ of the shirts over the weekend. She plans to split the remainder of the work equally over the next 5 evenings.

    a. What fraction of the shirts will Lenox iron each day after school?

    b. If Lenox has 40 shirts, how many shirts will she need to iron on Thursday and Friday?

7. Bo paid bills with $\dfrac{1}{2}$ of his paycheck and put $\dfrac{1}{5}$ of the remainder in savings. The rest of his paycheck he divided equally among the college accounts of his 3 children.

    a. What fraction of his paycheck went into each child's account?

    b. If Bo deposited $400 in each child's account, how much money was in Bo's original paycheck?

EUREKA
MATH™

©2015 Great Minds. eureka-math.org
G6-M2-SE-B1-1.3.1-01.2016

This page intentionally left blank

Student Edition

Eureka Math
Grade 6
Module 2

Special thanks go to the Gordan A. Cain Center and to the Department of Mathematics at Louisiana State University for their support in the development of *Eureka Math*.

For a free *Eureka Math* Teacher
Resource Pack, Parent Tip
Sheets, and more please
visit www.Eureka.tools

Published by the non-profit Great Minds

Copyright © 2015 Great Minds. No part of this work may be reproduced, sold, or commercialized, in whole or in part, without written permission from Great Minds. Non-commercial use is licensed pursuant to a Creative Commons Attribution-NonCommercial-ShareAlike 4.0 license; for more information, go to http://greatminds.net/maps/math/copyright. "Great Minds" and "Eureka Math" are registered trademarks of Great Minds.

Printed in the U.S.A.

This book may be purchased from the publisher at eureka-math.org

10 9 8 7 6 5 4 3

ISBN 978-1-63255-314-0

Lesson 2: Interpreting Division of a Whole Number by a Fraction—Visual Models

Classwork

Example 1

Question #_____

Write it as a division expression.   _____

Write it as a multiplication expression.   _____

Make a rough draft of a model to represent the problem:

©2015 Great Minds. eureka-math.org
G6-M2-SE-B1-1.3.1-01.2016

As you travel to each model, be sure to answer the following questions:

Original Question	Corresponding Division Expression	Corresponding Multiplication Expression	Write an Equation Showing the Equivalence of the Two Expressions.
1. How many $\frac{1}{2}$ miles are in 12 miles?			
2. How many quarter hours are in 5 hours?			
3. How many $\frac{1}{3}$ cups are in 9 cups?			
4. How many $\frac{1}{8}$ pizzas are in 4 pizzas?			
5. How many one-fifths are in 7 wholes?			

©2015 Great Minds. eureka-math.org
G6-M2-SE-B1-1.3.1-01.2016

Example 2

Molly has 9 cups of flour. If this is $\frac{3}{4}$ of the number she needs to make bread, how many cups does she need?

    a.   Construct the tape diagram by reading it backward. Draw a tape diagram and label the unknown.

    b.   Next, shade in $\frac{3}{4}$.

    c.   Label the shaded region to show that 9 is equal to $\frac{3}{4}$ of the total.

    d.   Analyze the model to determine the quotient.

Exercises 1–5

1. A construction company is setting up signs on 2 miles of road. If the company places a sign every $\frac{1}{4}$ mile, how many signs will it use?

2. George bought 4 submarine sandwiches for a birthday party. If each person will eat $\frac{2}{3}$ of a sandwich, how many people can George feed?

3. Miranda buys 6 pounds of nuts. If she puts $\frac{3}{4}$ pound in each bag, how many bags can she make?

EUREKA MATH™

4. Margo freezes 8 cups of strawberries. If this is $\frac{2}{3}$ of the total strawberries that she picked, how many cups of strawberries did Margo pick?

5. Regina is chopping up wood. She has chopped 10 logs so far. If the 10 logs represent $\frac{5}{8}$ of all the logs that need to be chopped, how many logs need to be chopped in all?

EUREKA
MATH™

©2015 Great Minds. eureka-math.org
G6-M2-SE-B1-1.3.1-01.2016

Problem Set

Rewrite each problem as a multiplication question. Model your answer.

1. Nicole used $\frac{3}{8}$ of her ribbon to wrap a present. If she used 6 feet of ribbon for the present, how much ribbon did Nicole have at first?

2. A Boy Scout has 3 meters of rope. He cuts the rope into cords $\frac{3}{5}$ m long. How many cords will he make?

3. 12 gallons of water fill a tank to $\frac{3}{4}$ capacity.

   a. What is the capacity of the tank?

   b. If the tank is then filled to capacity, how many half-gallon bottles can be filled with the water in the tank?

4. Hunter spent $\frac{2}{3}$ of his money on a video game before spending half of his remaining money on lunch. If his lunch costs $10, how much money did he have at first?

5. Students were surveyed about their favorite colors. $\frac{1}{4}$ of the students preferred red, $\frac{1}{8}$ of the students preferred blue, and $\frac{3}{5}$ of the remaining students preferred green. If 15 students preferred green, how many students were surveyed?

6. Mr. Scruggs got some money for his birthday. He spent $\frac{1}{5}$ of it on dog treats. Then, he divided the remainder equally among his 3 favorite charities.

   a. What fraction of his money did each charity receive?

   b. If he donated $60 to each charity, how much money did he receive for his birthday?

EUREKA
MATH™

©2015 Great Minds. eureka-math.org
G6-M2-SE-B1-1.3.1-01.2016

Lesson 3: Interpreting and Computing Division of a Fraction by a Fraction—More Models

Classwork

Opening Exercise

Draw a model to represent $12 \div 3$.

Create a question or word problem that matches your model.

Example 1

$$\frac{8}{9} \div \frac{2}{9}$$

Write the expression in unit form, and then draw a model to solve.

EUREKA MATH™

©2015 Great Minds. eureka-math.org
G6-M2-SE-B1-1.3.1-01.2016

Example 2

$$\frac{9}{12} \div \frac{3}{12}$$

Write the expression in unit form, and then draw a model to solve.

Example 3

$$\frac{7}{9} \div \frac{3}{9}$$

Write the expression in unit form, and then draw a model to solve.

Lesson 3:   Interpreting and Computing Division of a Fraction by a Fraction—More Models

©2015 Great Minds. eureka-math.org
G6-M2-SE-B1-1.3.1-01.2016

EUREKA MATH™

Exercises 1–6

Write an expression to represent each problem. Then, draw a model to solve.

1. How many fourths are in 3 fourths?

2. $\dfrac{4}{5} \div \dfrac{2}{5}$

EUREKA
MATH™

©2015 Great Minds. eureka-math.org
G6-M2-SE-B1-1.3.1-01.2016

3. $\dfrac{9}{4} \div \dfrac{3}{4}$

4. $\dfrac{7}{8} \div \dfrac{2}{8}$

5. $\dfrac{13}{10} \div \dfrac{2}{10}$

6. $\dfrac{11}{9} \div \dfrac{3}{9}$

Lesson 3:   Interpreting and Computing Division of a Fraction by a Fraction—More Models

©2015 Great Minds. eureka-math.org
G6-M2-SE-B1-1.3.1-01.2016

EUREKA
MATH™

Lesson Summary

When dividing a fraction by a fraction with the same denominator, we can use the general rule $\dfrac{a}{c} \div \dfrac{b}{c} = \dfrac{a}{b}$.

Problem Set

For the following exercises, rewrite the division expression in unit form. Then, find the quotient. Draw a model to support your answer.

1. $\dfrac{4}{5} \div \dfrac{1}{5}$

2. $\dfrac{8}{9} \div \dfrac{4}{9}$

3. $\dfrac{15}{4} \div \dfrac{3}{4}$

4. $\dfrac{13}{5} \div \dfrac{4}{5}$

Rewrite the expression in unit form, and find the quotient.

5. $\dfrac{10}{3} \div \dfrac{2}{3}$

6. $\dfrac{8}{5} \div \dfrac{3}{5}$

7. $\dfrac{12}{7} \div \dfrac{12}{7}$

Represent the division expression using unit form. Find the quotient. Show all necessary work.

8. A runner is $\dfrac{7}{8}$ mile from the finish line. If she can travel $\dfrac{3}{8}$ mile per minute, how long will it take her to finish the race?

9. An electrician has 4.1 meters of wire.

   a. How many strips $\dfrac{7}{10}$ m long can he cut?

   b. How much wire will he have left over?

10. Saeed bought $21\dfrac{1}{2}$ lb. of ground beef. He used $\dfrac{1}{4}$ of the beef to make tacos and $\dfrac{2}{3}$ of the remainder to make quarter-pound burgers. How many burgers did he make?

11. A baker bought some flour. He used $\dfrac{2}{5}$ of the flour to make bread and used the rest to make batches of muffins. If he used 16 lb. of flour making bread and $\dfrac{2}{3}$ lb. for each batch of muffins, how many batches of muffins did he make?

This page intentionally left blank

Lesson 4: Interpreting and Computing Division of a Fraction by a Fraction—More Models

Classwork

Opening Exercise

Write at least three equivalent fractions for each fraction below.

 a. $\dfrac{2}{3}$

 b. $\dfrac{10}{12}$

Example 1

Molly has $1\dfrac{3}{8}$ cups of strawberries. She needs $\dfrac{3}{8}$ cup of strawberries to make one batch of muffins. How many batches can Molly make?

Use a model to support your answer.

Example 2

Molly's friend, Xavier, also has $\frac{11}{8}$ cups of strawberries. He needs $\frac{3}{4}$ cup of strawberries to make a batch of tarts. How many batches can he make? Draw a model to support your solution.

Example 3

Find the quotient: $\frac{6}{8} \div \frac{2}{8}$. Use a model to show your answer.

©2015 Great Minds. eureka-math.org
G6-M2-SE-B1-1.3.1-01.2016

EUREKA
MATH™

Example 4

Find the quotient: $\dfrac{3}{4} \div \dfrac{2}{3}$. Use a model to show your answer.

Exercises 1–5

Find each quotient.

1. $\dfrac{6}{2} \div \dfrac{3}{4}$

EUREKA MATH™

Lesson 4:   Interpreting and Computing Division of a Fraction by a Fraction—More
Models

©2015 Great Minds. eureka-math.org
G6-M2-SE-B1-1.3.1-01.2016

S.21

2. $\dfrac{2}{3} \div \dfrac{2}{5}$

3. $\dfrac{7}{8} \div \dfrac{1}{2}$

4. $\dfrac{3}{5} \div \dfrac{1}{4}$

©2015 Great Minds. eureka-math.org
G6-M2-SE-B1-1.3.1-01.2016

EUREKA
MATH™

5. $\dfrac{5}{4} \div \dfrac{1}{3}$

©2015 Great Minds. eureka-math.org
G6-M2-SE-B1-1.3.1-01.2016

Problem Set

Calculate the quotient. If needed, draw a model.

1. $\dfrac{8}{9} \div \dfrac{4}{9}$

2. $\dfrac{9}{10} \div \dfrac{4}{10}$

3. $\dfrac{3}{5} \div \dfrac{1}{3}$

4. $\dfrac{3}{4} \div \dfrac{1}{5}$

©2015 Great Minds. eureka-math.org
G6-M2-SE-B1-1.3.1-01.2016

EUREKA MATH

Lesson 5: Creating Division Stories

Classwork

Opening Exercise

Tape Diagram:

$$\frac{8}{9} \div \frac{2}{9}$$

Number Line:

Molly's friend, Xavier, also has $\frac{11}{8}$ cups of strawberries. He needs $\frac{3}{4}$ cup of strawberries to make a batch of tarts. How many batches can he make? Draw a model to support your solution.

Example 1

$$\frac{1}{2} \div \frac{1}{8}$$

Step 1: Decide on an interpretation.

Step 2: Draw a model.

Step 3: Find the answer.

Step 4: Choose a unit.

Step 5: Set up a situation based upon the model.

EUREKA
MATH™

©2015 Great Minds. eureka-math.org
G6-M2-SE-B1-1.3.1-01.2016

Exercise 1

Using the same dividend and divisor, work with a partner to create your own story problem. You may use the same unit, but your situation must be unique. You could try another unit such as ounces, yards, or miles if you prefer.

Example 2

$$\frac{3}{4} \div \frac{1}{2}$$

Step 1: Decide on an interpretation.

Step 2: Draw a diagram.

Step 3: Find the answer.

Step 4: Choose a unit.

Step 5: Set up a situation based on the model.

Exercise 2

Using the same dividend and divisor, work with a partner to create your own story problem. You may use the same unit, but your situation must be unique. You could try another unit such as cups, yards, or miles if you prefer.

Lesson 5:   Creating Division Stories

©2015 Great Minds. eureka-math.org
G6-M2-SE-B1-1.3.1-01.2016

Lesson Summary

The method of creating division stories includes five steps:

Step 1: Decide on an interpretation (measurement or partitive). Today we used measurement division.

Step 2: Draw a model.

Step 3: Find the answer.

Step 4: Choose a unit.

Step 5: Set up a situation based on the model. This means writing a story problem that is interesting, realistic, and short. It may take several attempts before you find a story that works well with the given dividend and divisor.

Problem Set

Solve.

1. How many sixteenths are in $\frac{15}{16}$?

2. How many $\frac{1}{4}$ teaspoon doses are in $\frac{7}{8}$ teaspoon of medicine?

3. How many $\frac{2}{3}$ cups servings are in a 4 cup container of food?

4. Write a measurement division story problem for $6 \div \frac{3}{4}$.

5. Write a measurement division story problem for $\frac{5}{12} \div \frac{1}{6}$.

6. Fill in the blank to complete the equation. Then, find the quotient and draw a model to support your solution.

   a. $\frac{1}{2} \div 5 = \frac{1}{\square}$ of $\frac{1}{2}$

   b. $\frac{3}{4} \div 6 = \frac{1}{\square}$ of $\frac{3}{4}$

7. $\frac{4}{5}$ of the money collected from a fundraiser was divided equally among 8 grades. What fraction of the money did each grade receive?

8. Meyer used 6 loads of gravel to cover $\frac{2}{5}$ of his driveway. How many loads of gravel will he need to cover his entire driveway?

9. An athlete plans to run 3 miles. Each lap around the school yard is $\frac{3}{7}$ mile. How many laps will the athlete run?

10. Parks spent $\frac{1}{3}$ of his money on a sweater. He spent $\frac{3}{5}$ of the remainder on a pair of jeans. If he has $36 left, how much did the sweater cost?

EUREKA MATH

©2015 Great Minds. eureka-math.org
G6-M2-SE-B1-1.3.1-01.2016

Lesson 6: More Division Stories

Classwork

Example 1

Divide $50 \div \frac{2}{3}$.

Step 1: Decide on an interpretation.

Step 2: Draw a model.

Step 3: Find the answer.

Step 4: Choose a unit.

©2015 Great Minds. eureka-math.org
G6-M2-SE-B1-1.3.1-01.2016

Step 5: Set up a situation based upon the model.

Exercise 1

Using the same dividend and divisor, work with a partner to create your own story problem. You may use the same unit, dollars, but your situation must be unique. You could try another unit, such as miles, if you prefer.

Example 2

Divide $\frac{1}{2} \div \frac{3}{4}$.

Step 1: Decide on an interpretation.

Step 2: Draw a model.

EUREKA
MATH™

©2015 Great Minds. eureka-math.org
G6-M2-SE-B1-1.3.1-01.2016

Step 3: Find the answer.

Step 4: Choose a unit.

Step 5: Set up a situation based upon the model.

Exercise 2

Using the same dividend and divisor, work with a partner to create your own story problem. Try a different unit.

Problem Set

Solve.

1. $\frac{15}{16}$ is 1 sixteenth groups of what size?

2. $\frac{7}{8}$ teaspoons is $\frac{1}{4}$ groups of what size?

3. A 4-cup container of food is $\frac{2}{3}$ groups of what size?

4. Write a partitive division story problem for $6 \div \frac{3}{4}$.

5. Write a partitive division story problem for $\frac{5}{12} \div \frac{1}{6}$.

6. Fill in the blank to complete the equation. Then, find the quotient, and draw a model to support your solution.

    a. $\frac{1}{4} \div 7 = \frac{1}{\square}$ of $\frac{1}{4}$

    b. $\frac{5}{6} \div 4 = \frac{1}{\square}$ of $\frac{5}{6}$

7. There is $\frac{3}{5}$ of a pie left. If 4 friends wanted to share the pie equally, how much would each friend receive?

8. In two hours, Holden completed $\frac{3}{4}$ of his race. How long will it take Holden to complete the entire race?

9. Sam cleaned $\frac{1}{3}$ of his house in 50 minutes. How many hours will it take him to clean his entire house?

10. It took Mario 10 months to beat $\frac{5}{8}$ of the levels on his new video game. How many years will it take for Mario to beat all the levels?

11. A recipe calls for $1\frac{1}{2}$ cups of sugar. Marley only has measuring cups that measure $\frac{1}{4}$ cup. How many times will Marley have to fill the measuring cup?

©2015 Great Minds. eureka-math.org
G6-M2-SE-B1-1.3.1-01.2016

EUREKA
MATH™

Lesson 7: The Relationship Between Visual Fraction Models and Equations

Classwork

Example 1

Model the following using a partitive interpretation.

$$\frac{3}{4} \div \frac{2}{5}$$

Shade 2 of the 5 sections $\left(\frac{2}{5}\right)$.

Label the part that is known $\left(\frac{3}{4}\right)$.

Make notes below on the math sentences needed to solve the problem.

EUREKA MATH™

©2015 Great Minds. eureka-math.org
G6-M2-SE-B1-1.3.1-01.2016

Example 2

Model the following using a measurement interpretation.

$$\frac{3}{5} \div \frac{1}{4}$$

Example 3

$$\frac{2}{3} \div \frac{3}{4}$$

Show the number sentences below.

©2015 Great Minds. eureka-math.org
G6-M2-SE-B1-1.3.1-01.2016

Lesson Summary

Connecting models of fraction division to multiplication through the use of reciprocals helps in understanding the *invert and multiply* rule. That is, given two fractions $\frac{a}{b}$ and $\frac{c}{d}$, we have the following:

$$\frac{a}{b} \div \frac{c}{d} = \frac{a}{b} \times \frac{d}{c}.$$

Problem Set

Invert and multiply to divide.

1.

   a. $\frac{2}{3} \div \frac{1}{4}$ b. $\frac{2}{3} \div 4$ c. $4 \div \frac{2}{3}$

2.

   a. $\frac{1}{3} \div \frac{1}{4}$ b. $\frac{1}{8} \div \frac{3}{4}$ c. $\frac{9}{4} \div \frac{6}{5}$

3.

   a. $\frac{2}{3} \div \frac{3}{4}$ b. $\frac{3}{5} \div \frac{3}{2}$ c. $\frac{22}{4} \div \frac{2}{5}$

4. Summer used $\frac{2}{5}$ of her ground beef to make burgers. If she used $\frac{3}{4}$ pounds of beef, how much beef did she have at first?

5. Alistair has 5 half-pound chocolate bars. It takes $1\frac{1}{2}$ pounds of chocolate, broken into chunks, to make a batch of cookies. How many batches can Alistair make with the chocolate he has on hand?

6. Draw a model that shows $\frac{2}{5} \div \frac{1}{3}$. Find the answer as well.

7. Draw a model that shows $\frac{3}{4} \div \frac{1}{2}$. Find the answer as well.

©2015 Great Minds. eureka-math.org
G6-M2-SE-B1-1.3.1-01.2016

This page intentionally left blank

Lesson 8: Dividing Fractions and Mixed Numbers

Classwork

Example 1: Introduction to Calculating the Quotient of a Mixed Number and a Fraction

a. Carli has $4\frac{1}{2}$ walls left to paint in order for all the bedrooms in her house to have the same color paint.

However, she has used almost all of her paint and only has $\frac{5}{6}$ of a gallon left.

How much paint can she use on each wall in order to have enough to paint the remaining walls?

b. Calculate the quotient.

$$\frac{2}{5} \div 3\frac{4}{7}$$

©2015 Great Minds. eureka-math.org
G6-M2-SE-B1-1.3.1-01.2016

Exercise

Show your work for the memory game in the boxes provided below.

| A. |
| B. |
| C. |
| D. |
| E. |
| F. |
| G. |
| H. |
| I. |
| J. |
| K. |
| L. |

©2015 Great Minds. eureka-math.org
G6-M2-SE-B1-1.3.1-01.2016

Problem Set

Calculate each quotient.

1. $\dfrac{2}{5} \div 3\dfrac{1}{10}$

2. $4\dfrac{1}{3} \div \dfrac{4}{7}$

3. $3\dfrac{1}{6} \div \dfrac{9}{10}$

4. $\dfrac{5}{8} \div 2\dfrac{7}{12}$

©2015 Great Minds. eureka-math.org
G6-M2-SE-B1-1.3.1-01.2016

This page intentionally left blank

Lesson 9: Sums and Differences of Decimals

Classwork

Example 1

$$25\frac{3}{10} + 376\frac{77}{100}$$

Example 2

$$426\frac{1}{5} - 275\frac{1}{2}$$

©2015 Great Minds. eureka-math.org
G6-M2-SE-B1-1.3.1-01.2016

Exercises

Calculate each sum or difference.

1. Samantha and her friends are going on a road trip that is $245\frac{7}{50}$ miles long. They have already driven $128\frac{53}{100}$.
   How much farther do they have to drive?

2. Ben needs to replace two sides of his fence. One side is $367\frac{9}{100}$ meters long, and the other is $329\frac{3}{10}$ meters long.
   How much fence does Ben need to buy?

3. Mike wants to paint his new office with two different colors. If he needs $4\frac{4}{5}$ gallons of red paint and $3\frac{1}{10}$ gallons of
   brown paint, how much paint does he need in total?

©2015 Great Minds. eureka-math.org
G6-M2-SE-B1-1.3.1-01.2016

4. After Arianna completed some work, she figured she still had $78\frac{21}{100}$ pictures to paint. If she completed another $34\frac{23}{25}$ pictures, how many pictures does Arianna still have to paint?

Use a calculator to convert the fractions into decimals before calculating the sum or difference.

5. Rahzel wants to determine how much gasoline he and his wife use in a month. He calculated that he used $78\frac{1}{3}$ gallons of gas last month. Rahzel's wife used $41\frac{3}{8}$ gallons of gas last month. How much total gas did Rahzel and his wife use last month? Round your answer to the nearest hundredth.

Problem Set

1. Find each sum or difference.

   a. $381\frac{1}{10} - 214\frac{43}{100}$

   b. $32\frac{3}{4} - 12\frac{1}{2}$

   c. $517\frac{37}{50} + 312\frac{3}{100}$

   d. $632\frac{16}{25} + 32\frac{3}{10}$

   e. $421\frac{3}{50} - 212\frac{9}{10}$

2. Use a calculator to find each sum or difference. Round your answer to the nearest hundredth.

   a. $422\frac{3}{7} - 367\frac{5}{9}$

   b. $23\frac{1}{5} + 45\frac{7}{8}$

©2015 Great Minds. eureka-math.org
G6-M2-SE-B1-1.3.1-01.2016

EUREKA MATH™

Lesson 10: The Distributive Property and the Products of Decimals

Classwork

Opening Exercise

Calculate the product.

    a. 200×32.6

    b. 500×22.12

Example 1: Introduction to Partial Products

Use partial products and the distributive property to calculate the product.

200×32.6

Example 2: Introduction to Partial Products

Use partial products and the distributive property to calculate the area of the rectangular patio shown below.

22.12 ft.

500 ft.

Exercises

Use the boxes below to show your work for each station. Make sure that you are putting the solution for each station in the correct box.

Station One:

Station Two:

Station Three:

Station Four:

Station Five:

©2015 Great Minds. eureka-math.org
G6-M2-SE-B1-1.3.1-01.2016

Problem Set

Calculate the product using partial products.

1. 400×45.2

2. 14.9×100

3. 200×38.4

4. 900×20.7

5. 76.2×200

This page intentionally left blank

Lesson 11: Fraction Multiplication and the Products of Decimals

Classwork

Exploratory Challenge

You not only need to solve each problem, but your groups also need to prove to the class that the decimal in the product is located in the correct place. As a group, you are expected to present your informal proof to the class.

a. Calculate the product. 34.62×12.8

b. Xavier earns $11.50 per hour working at the nearby grocery store. Last week, Xavier worked for 13.5 hours. How much money did Xavier earn last week? Remember to round to the nearest penny.

Discussion

Record notes from the Discussion in the box below.

Exercises

1. Calculate the product. 324.56×54.82

2. Kevin spends \$11.25 on lunch every week during the school year. If there are 35.5 weeks during the school year, how much does Kevin spend on lunch over the entire school year? Remember to round to the nearest penny.

©2015 Great Minds. eureka-math.org
G6-M2-SE-B1-1.3.1-01.2016

EUREKA
MATH

3. Gunnar's car gets 22.4 miles per gallon, and his gas tank can hold 17.82 gallons of gas. How many miles can Gunnar travel if he uses all of the gas in the gas tank?

4. The principal of East High School wants to buy a new cover for the sand pit used in the long-jump competition. He measured the sand pit and found that the length is 29.2 feet and the width is 9.8 feet. What will the area of the new cover be?

Problem Set

Solve each problem. Remember to round to the nearest penny when necessary.

1. Calculate the product. 45.67×32.58

2. Deprina buys a large cup of coffee for $4.70 on her way to work every day. If there are 24 workdays in the month, how much does Deprina spend on coffee throughout the entire month?

3. Krego earns $2,456.75 every month. He also earns an extra $4.75 every time he sells a new gym membership. Last month, Krego sold 32 new gym memberships. How much money did Krego earn last month?

4. Kendra just bought a new house and needs to buy new sod for her backyard. If the dimensions of her yard are 24.6 feet by 14.8 feet, what is the area of her yard?

©2015 Great Minds. eureka-math.org
G6-M2-SE-B1-1.3.1-01.2016

Lesson 12: Estimating Digits in a Quotient

Classwork

Discussion

Divide 150 by 30.

Exercises 1–5

Round to estimate the quotient. Then, compute the quotient using a calculator, and compare the estimation to the quotient.

1.   2,970 ÷ 11

   a.   Round to a one-digit arithmetic fact. Estimate the quotient.

   b.   Use a calculator to find the quotient. Compare the quotient to the estimate.

2. $4,752 \div 12$

   a.   Round to a one-digit arithmetic fact. Estimate the quotient.

   b.   Use a calculator to find the quotient. Compare the quotient to the estimate.

3. $11,647 \div 19$

   a.   Round to a one-digit arithmetic fact. Estimate the quotient.

   b.   Use a calculator to find the quotient. Compare the quotient to the estimate.

Lesson 12:   Estimating Digits in a Quotient

EUREKA MATH™

©2015 Great Minds. eureka-math.org
G6-M2-SE-B1-1.3.1-01.2016

4. $40,644 \div 18$

    a.   Round to a one-digit arithmetic fact. Estimate the quotient.

    b.   Use a calculator to find the quotient. Compare the quotient to the estimate.

5. $49,170 \div 15$

    a.   Round to a one-digit arithmetic fact. Estimate the quotient.

    b.   Use a calculator to find the quotient. Compare the quotient to the estimate.

Example 3: Extend Estimation and Place Value to the Division Algorithm

Estimate and apply the division algorithm to evaluate the expression $918 \div 27$.

©2015 Great Minds. eureka-math.org
G6-M2-SE-B1-1.3.1-01.2016

EUREKA
MATH™

Problem Set

Round to estimate the quotient. Then, compute the quotient using a calculator, and compare the estimate to the quotient.

1. $715 \div 11$

2. $7,884 \div 12$

3. $9,646 \div 13$

4. $11,942 \div 14$

5. $48,825 \div 15$

6. $135,296 \div 16$

7. $199,988 \div 17$

8. $116,478 \div 18$

9. $99,066 \div 19$

10. $181,800 \div 20$

©2015 Great Minds. eureka-math.org
G6-M2-SE-B1-1.3.1-01.2016

This page intentionally left blank

Lesson 13: Dividing Multi-Digit Numbers Using the Algorithm

Classwork

Example 1

Divide $70{,}072 \div 19$.

   a.   Estimate:

   b.   Create a table to show the multiples of 19.

Multiples of 19

c.   Use the algorithm to divide $70,072 \div 19$. Check your work.

$$1\ 9\ \overline{\smash{)}\ 7\quad 0\quad 0\quad 7\quad 2}$$

Example 2

Divide $14,175 \div 315$.

   a.   Estimate:

   b.   Use the algorithm to divide $14,175 \div 315$. Check your work.

©2015 Great Minds. eureka-math.org
G6-M2-SE-B1-1.3.1-01.2016

EUREKA
MATH™

Exercises 1–5

For each exercise,

    a.   Estimate.

    b.   Divide using the algorithm, explaining your work using place value.

1. $484{,}692 \div 78$

    a.   Estimate:

    b.

2. $281{,}886 \div 33$

    a.   Estimate:

    b.

EUREKA MATH™

©2015 Great Minds. eureka-math.org
G6-M2-SE-B1-1.3.1-01.2016

3. 2,295,517 ÷ 37
    a. Estimate:

    b.

4. 952,448 ÷ 112
    a. Estimate:

    b.

EUREKA
MATH™

©2015 Great Minds. eureka-math.org
G6-M2-SE-B1-1.3.1-01.2016

5. 1,823,535 ÷ 245

   a. Estimate:

   b.

Problem Set

Divide using the division algorithm.

1. $1,634 \div 19$

2. $2,450 \div 25$

3. $22,274 \div 37$

4. $21,361 \div 41$

5. $34,874 \div 53$

6. $50,902 \div 62$

7. $70,434 \div 78$

8. $91,047 \div 89$

9. $115,785 \div 93$

10. $207,968 \div 97$

11. $7,735 \div 119$

12. $21,948 \div 354$

13. $72,372 \div 111$

14. $74,152 \div 124$

15. $182,727 \div 257$

16. $396,256 \div 488$

17. $730,730 \div 715$

18. $1,434,342 \div 923$

19. $1,775,296 \div 32$

20. $1,144,932 \div 12$

©2015 Great Minds. eureka-math.org
G6-M2-SE-B1-1.3.1-01.2016

Lesson 14: The Division Algorithm—Converting Decimal Division into Whole Number Division Using Fractions

Classwork

Opening Exercise

Divide $\frac{1}{2} \div \frac{1}{10}$. Use a tape diagram to support your reasoning.

Relate the model to the invert and multiply rule.

©2015 Great Minds. eureka-math.org
G6-M2-SE-B1-1.3.1-01.2016

Example 1

Evaluate the expression. Use a tape diagram to support your answer.

$0.5 \div 0.1$

Rewrite $0.5 \div 0.1$ as a fraction.

Express the divisor as a whole number.

Exercises 1–3

Convert the decimal division expressions to fractional division expressions in order to create whole number divisors. You do not need to find the quotients. Explain the movement of the decimal point. The first exercise has been completed for you.

1. $18.6 \div 2.3$

    $$\frac{18.6}{2.3} \times \frac{10}{10} = \frac{186}{23}$$

    $186 \div 23$

    I multiplied both the dividend and the divisor by ten, or by one power of ten, so each decimal point moved one place to the right because they grew larger by ten.

2. $14.04 \div 4.68$

©2015 Great Minds. eureka-math.org
G6-M2-SE-B1-1.3.1-01.2016

EUREKA
MATH™

3. $0.162 \div 0.036$

Example 2

Evaluate the expression. First, convert the decimal division expression to a fractional division expression in order to create a whole number divisor.

$25.2 \div 0.72$

Use the division algorithm to find the quotient.

Exercises 4–7

Convert the decimal division expressions to fractional division expressions in order to create whole number divisors. Compute the quotients using the division algorithm. Check your work with a calculator.

4. $2,000 \div 3.2$

5. $3,581.9 \div 4.9$

Lesson 14:   The Division Algorithm—Converting Decimal Division into Whole Number
                        Division Using Fractions

©2015 Great Minds. eureka-math.org
G6-M2-SE-B1-1.3.1-01.2016

6. $893.76 \div 0.21$

7. $6.194 \div 0.326$

EUREKA
MATH™

©2015 Great Minds. eureka-math.org
G6-M2-SE-B1-1.3.1-01.2016

Example 3

A plane travels 3,625.26 miles in 6.9 hours. What is the plane's unit rate?

Represent this situation with a fraction.

Represent this situation using the same units.

Estimate the quotient.

Express the divisor as a whole number.

Use the division algorithm to find the quotient.

Use multiplication to check your work.

Lesson 14:   The Division Algorithm—Converting Decimal Division into Whole Number
Division Using Fractions

©2015 Great Minds. eureka-math.org
G6-M2-SE-B1-1.3.1-01.2016

EUREKA MATH™

Problem Set

Convert decimal division expressions to fractional division expressions to create whole number divisors.

1. $35.7 \div 0.07$

2. $486.12 \div 0.6$

3. $3.43 \div 0.035$

4. $5{,}418.54 \div 0.009$

5. $812.5 \div 1.25$

6. $17.343 \div 36.9$

Estimate quotients. Convert decimal division expressions to fractional division expressions to create whole number divisors. Compute the quotients using the division algorithm. Check your work with a calculator and your estimates.

7. Norman purchased 3.5 lb. of his favorite mixture of dried fruits to use in a trail mix. The total cost was $16.87. How much does the fruit cost per pound?

8. Divide: $994.14 \div 18.9$

9. Daryl spent $4.68 on each pound of trail mix. He spent a total of $14.04. How many pounds of trail mix did he purchase?

10. Mamie saved $161.25. This is 25% of the amount she needs to save. How much money does Mamie need to save?

11. Kareem purchased several packs of gum to place in gift baskets for $1.26 each. He spent a total of $8.82. How many packs of gum did he buy?

12. Jerod is making candles from beeswax. He has 132.72 ounces of beeswax. If each candle uses 8.4 ounces of beeswax, how many candles can he make? Will there be any wax left over?

13. There are 20.5 cups of batter in the bowl. This represents 0.4 of the entire amount of batter needed for a recipe. How many cups of batter are needed?

14. Divide: $159.12 \div 6.8$

15. Divide: $167.67 \div 8.1$

This page intentionally left blank

Lesson 15: The Division Algorithm—Converting Decimal Division into Whole Number Division Using Mental Math

Classwork

Opening Exercise

Use mental math to evaluate the numeric expressions.

    a. $99 + 44$

    b. $86 - 39$

    c. 50×14

    d. $180 \div 5$

Example 1: Use Mental Math to Find Quotients

Use mental math to evaluate $105 \div 35$.

Exercises 1–4

Use mental math techniques to evaluate the expressions.

1. $770 \div 14$

2. $1{,}005 \div 5$

3. $1{,}500 \div 8$

4. $1{,}260 \div 5$

Lesson 15:   The Division Algorithm—Converting Decimal Division into Whole Number Division Using Mental Math

©2015 Great Minds. eureka-math.org
G6-M2-SE-B1-1.3.1-01.2016

Example 2: Mental Math and Division of Decimals

Evaluate the expression $175 \div 3.5$ using mental math techniques.

Exercises 5–7

Use mental math techniques to evaluate the expressions.

5. $25 \div 6.25$

6. $6.3 \div 1.5$

7. $425 \div 2.5$

EUREKA MATH

Lesson 15:   The Division Algorithm—Converting Decimal Division into Whole
Number Division Using Mental Math

S.77

©2015 Great Minds. eureka-math.org
G6-M2-SE-B1-1.3.1-01.2016

Example 3: Mental Math and the Division Algorithm

Evaluate the expression 4,564 ÷ 3.5 using mental math techniques and the division algorithm.

©2015 Great Minds. eureka-math.org
G6-M2-SE-B1-1.3.1-01.2016

Example 4: Mental Math and Reasonable Work

Shelly was given this number sentence and was asked to place the decimal point correctly in the quotient.

$$55.6875 \div 6.75 = 0.825$$

Do you agree with Shelly?

Divide to prove your answer is correct.

Problem Set

Use mental math, estimation, and the division algorithm to evaluate the expressions.

1. $118.4 \div 6.4$

2. $314.944 \div 3.7$

3. $1,840.5072 \div 23.56$

4. $325 \div 2.5$

5. $196 \div 3.5$

6. $405 \div 4.5$

7. $3,437.5 \div 5.5$

8. $393.75 \div 5.25$

9. $2,625 \div 6.25$

10. $231 \div 8.25$

11. $92 \div 5.75$

12. $196 \div 12.25$

13. $117 \div 6.5$

14. $936 \div 9.75$

15. $305 \div 12.2$

Place the decimal point in the correct place to make the number sentence true.

16. $83.375 \div 2.3 = 3,625$

17. $183.575 \div 5,245 = 3.5$

18. $326,025 \div 9.45 = 3.45$

19. $449.5 \div 725 = 6.2$

20. $446,642 \div 85.4 = 52.3$

©2015 Great Minds. eureka-math.org
G6-M2-SE-B1-1.3.1-01.2016

EUREKA MATH™

Lesson 16: Even and Odd Numbers

Classwork

Opening Exercise

   a.   What is an even number?

   b.   List some examples of even numbers.

   c.   What is an odd number?

   d.   List some examples of odd numbers.

What happens when we add two even numbers? Do we always get an even number?

©2015 Great Minds. eureka-math.org
G6-M2-SE-B1-1.3.1-01.2016

Exercises 1–3

1. Why is the sum of two even numbers even?

    a. Think of the problem $12 + 14$. Draw dots to represent each number.

    b. Circle pairs of dots to determine if any of the dots are left over.

    c. Is this true every time two even numbers are added together? Why or why not?

2. Why is the sum of two odd numbers even?

    a. Think of the problem $11 + 15$. Draw dots to represent each number.

    b. Circle pairs of dots to determine if any of the dots are left over.

    c. Is this true every time two odd numbers are added together? Why or why not?

©2015 Great Minds. eureka-math.org
G6-M2-SE-B1-1.3.1-01.2016

EUREKA MATH™

3. Why is the sum of an even number and an odd number odd?

   a. Think of the problem $14 + 11$. Draw dots to represent each number.

   b. Circle pairs of dots to determine if any of the dots are left over.

   c. Is this true every time an even number and an odd number are added together? Why or why not?

   d. What if the first addend is odd and the second is even? Is the sum still odd? Why or why not? For example, if we had $11 + 14$, would the sum be odd?

Let's sum it up:

   ▪

   ▪

   ▪

EUREKA
MATH™

©2015 Great Minds. eureka-math.org
G6-M2-SE-B1-1.3.1-01.2016

Exploratory Challenge/Exercises 4–6

4. The product of two even numbers is even.

5. The product of two odd numbers is odd.

6. The product of an even number and an odd number is even.

©2015 Great Minds. eureka-math.org
G6-M2-SE-B1-1.3.1-01.2016

EUREKA
MATH™

Lesson Summary

Adding:

- The sum of two even numbers is even.
- The sum of two odd numbers is even.
- The sum of an even number and an odd number is odd.

Multiplying:

- The product of two even numbers is even.
- The product of two odd numbers is odd.
- The product of an even number and an odd number is even.

Problem Set

Without solving, tell whether each sum or product is even or odd. Explain your reasoning.

1. $346 + 721$

2. $4,690 \times 141$

3. $1,462,891 \times 745,629$

4. $425,922 + 32,481,064$

5. $32 + 45 + 67 + 91 + 34 + 56$

This page intentionally left blank

Lesson 17: Divisibility Tests for 3 and 9

Classwork

Opening Exercise

Below is a list of 10 numbers. Place each number in the circle(s) that is a factor of the number. Some numbers can be placed in more than one circle. For example, if 32 were on the list, it would be placed in the circles with 2, 4, and 8 because they are all factors of 32.

24; 36; 80; 115; 214; 360; 975; 4,678; 29,785; 414,940

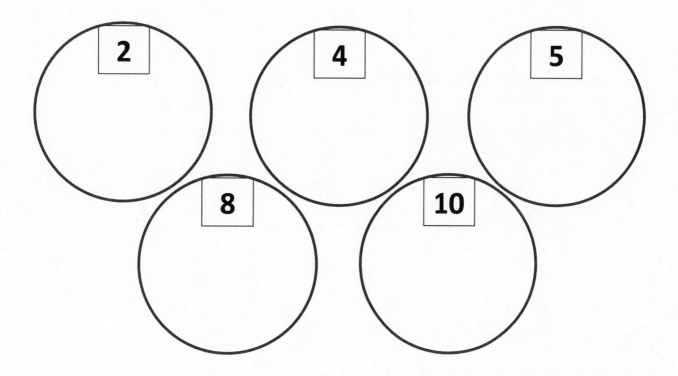

EUREKA MATH™

©2015 Great Minds. eureka-math.org
G6-M2-SE-B1-1.3.1-01.2016

Discussion

- Divisibility rule for 2:

- Divisibility rule for 4:

- Divisibility rule for 5:

- Divisibility rule for 8:

- Divisibility rule for 10:

- Decimal numbers with fraction parts do not follow the divisibility tests.

- Divisibility rule for 3:

- Divisibility rule for 9:

This example shows how to apply the two new divisibility rules we just discussed.

Explain why 378 is divisible by 3 and 9.

   a. Expand 378.

©2015 Great Minds. eureka-math.org
G6-M2-SE-B1-1.3.1-01.2016

b.   Decompose the expression to factor by 9.

c.   Factor the 9.

d.   What is the sum of the three digits?

e.   Is 18 divisble by 9?

f.   Is the number 378 divisible by 9? Why or why not?

g.   Is the number 378 divisible by 3? Why or why not?

Example 2

Is 3,822 divisible by 3 or 9? Why or why not?

Exercises 1–5

Circle ALL the numbers that are factors of the given number. Complete any necessary work in the space provided.

1. 2,838 is divisible by

    3

    9

    4

    Explain your reasoning for your choice(s).

2. 34,515 is divisible by

    3

    9

    5

    Explain your reasoning for your choice(s).

3. 10,534,341 is divisible by

    3

    9

    2

    Explain your reasoning for your choice(s).

©2015 Great Minds. eureka-math.org
G6-M2-SE-B1-1.3.1-01.2016

EUREKA
MATH

4. 4,320 is divisible by

    3

    9

    10

    Explain your reasoning for your choice(s).

5. 6,240 is divisible by

    3

    9

    8

    Explain your reasoning for your choice(s).

©2015 Great Minds. eureka-math.org
G6-M2-SE-B1-1.3.1-01.2016

Lesson Summary

To determine if a number is divisible by 3 or 9:

- Calculate the sum of the digits.
- If the sum of the digits is divisible by 3, the entire number is divisible by 3.
- If the sum of the digits is divisible by 9, the entire number is divisible by 9.

Note: If a number is divisible by 9, the number is also divisible by 3.

Problem Set

1. Is 32,643 divisible by both 3 and 9? Why or why not?

2. Circle all the factors of 424,380 from the list below.

   2   3   4   5   8   9   10

3. Circle all the factors of 322,875 from the list below.

   2   3   4   5   8   9   10

4. Write a 3-digit number that is divisible by both 3 and 4. Explain how you know this number is divisible by 3 and 4.

5. Write a 4-digit number that is divisible by both 5 and 9. Explain how you know this number is divisible by 5 and 9.

EUREKA
MATH™

©2015 Great Minds. eureka-math.org
G6-M2-SE-B1-1.3.1-01.2016

Lesson 18: Least Common Multiple and Greatest Common Factor

Classwork

Opening

The *greatest common factor* of two whole numbers (not both zero) is the greatest whole number that is a factor of each number. The greatest common factor of two whole numbers a and b is denoted by GCF (a, b).

The *least common multiple* of two whole numbers is the smallest whole number greater than zero that is a multiple of each number. The least common multiple of two whole numbers a and b is denoted by LCM (a, b).

Example 1: Greatest Common Factor

Find the greatest common factor of 12 and 18.

- Listing these factor pairs in order helps ensure that no common factors are missed. Start with 1 multiplied by the number.
- Circle all factors that appear on both lists.
- Place a triangle around the greatest of these common factors.

GCF $(12, 18)$

12

18

©2015 Great Minds. eureka-math.org
G6-M2-SE-B1-1.3.1-01.2016

Example 2: Least Common Multiple

Find the least common multiple of 12 and 18.

LCM (12, 18)

Write the first 10 multiples of 12.

Write the first 10 multiples of 18.

Circle the multiples that appear on both lists.

Put a rectangle around the least of these common multiples.

Exercises

Station 1: Factors and GCF

Choose one of these problems that has not yet been solved. Solve it together on your student page. Then, use your marker to copy your work neatly on the chart paper. Use your marker to cross out your choice so that the next group solves a different problem.

GCF (30, 50)

GCF (30, 45)

GCF (45, 60)

GCF (42, 70)

GCF (96, 144)

Lesson 18:   Least Common Multiple and Greatest Common Factor

©2015 Great Minds. eureka-math.org
G6-M2-SE-B1-1.3.1-01.2016

Next, choose one of these problems that has not yet been solved:

a. There are 18 girls and 24 boys who want to participate in a Trivia Challenge. If each team must have the same ratio of girls and boys, what is the greatest number of teams that can enter? Find how many boys and girls each team would have.

b. Ski Club members are preparing identical welcome kits for new skiers. The Ski Club has 60 hand-warmer packets and 48 foot-warmer packets. Find the greatest number of identical kits they can prepare using all of the hand-warmer and foot-warmer packets. How many hand-warmer packets and foot-warmer packets would each welcome kit have?

c. There are 435 representatives and 100 senators serving in the United States Congress. How many identical groups with the same numbers of representatives and senators could be formed from all of Congress if we want the largest groups possible? How many representatives and senators would be in each group?

d. Is the GCF of a pair of numbers ever equal to one of the numbers? Explain with an example.

e. Is the GCF of a pair of numbers ever greater than both numbers? Explain with an example.

©2015 Great Minds. eureka-math.org
G6-M2-SE-B1-1.3.1-01.2016

This page intentionally left blank

Station 2: Multiples and LCM

Choose one of these problems that has not yet been solved. Solve it together on your student page. Then, use your marker to copy your work neatly on the chart paper. Use your marker to cross out your choice so that the next group solves a different problem.

LCM (9, 12)

LCM (8, 18)

LCM (4, 30)

LCM (12, 30)

LCM (20, 50)

Next, choose one of these problems that has not yet been solved. Solve it together on your student page. Then, use your marker to copy your work neatly on this chart paper and to cross out your choice so that the next group solves a different problem.

a. Hot dogs come packed 10 in a package. Hot dog buns come packed 8 in a package. If we want one hot dog for each bun for a picnic with none left over, what is the least amount of each we need to buy? How many packages of each item would we have to buy?

b. Starting at 6:00 a.m., a bus stops at my street corner every 15 minutes. Also starting at 6:00 a.m., a taxi cab comes by every 12 minutes. What is the next time both a bus and a taxi are at the corner at the same time?

c. Two gears in a machine are aligned by a mark drawn from the center of one gear to the center of the other. If the first gear has 24 teeth, and the second gear has 40 teeth, how many revolutions of the first gear are needed until the marks line up again?

©2015 Great Minds. eureka-math.org
G6-M2-SE-B1-1.3.1-01.2016

This page intentionally left blank

d. Is the LCM of a pair of numbers ever equal to one of the numbers? Explain with an example.

e. Is the LCM of a pair of numbers ever less than both numbers? Explain with an example.

Station 3: Using Prime Factors to Determine GCF

Choose one of these problems that has not yet been solved. Solve it together on your student page. Then, use your marker to copy your work neatly on the chart paper and to cross out your choice so that the next group solves a different problem.

GCF (30, 50)

GCF (30, 45)

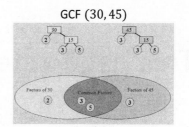

GCF (45, 60)

GCF (42, 70)

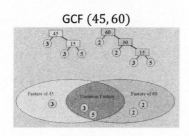
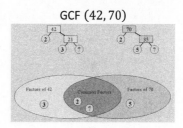

GCF (96, 144)

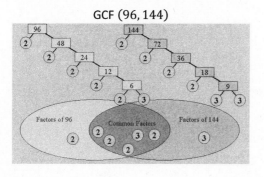

EUREKA
MATH™

©2015 Great Minds. eureka-math.org
G6-M2-SE-B1-1.3.1-01.2016

This page intentionally left blank

Next, choose one of these problems that has not yet been solved:

a. Would you rather find all the factors of a number or find all the prime factors of a number? Why?

b. Find the GCF of your original pair of numbers.

c. Is the product of your LCM and GCF less than, greater than, or equal to the product of your numbers?

d. Glenn's favorite number is very special because it reminds him of the day his daughter, Sarah, was born. The factors of this number do not repeat, and all the prime numbers are less than 12. What is Glenn's number? When was Sarah born?

Station 4: Applying Factors to the Distributive Property

Choose one of these problems that has not yet been solved. Solve it together on your student page. Then, use your marker to copy your work neatly on the chart paper and to cross out your choice so that the next group solves a different problem.

Find the GCF from the two numbers, and rewrite the sum using the distributive property.

1. $12 + 18 =$

2. $42 + 14 =$

3. $36 + 27 =$

4. $16 + 72 =$

5. $44 + 33 =$

This page intentionally left blank

Next, add another example to one of these two statements applying factors to the distributive property.

Choose any numbers for n, a, and b.

$n(a) + n(b) = n(a + b)$

$n(a) - n(b) = n(a - b)$

Problem Set

Complete the remaining stations from class.

©2015 Great Minds. eureka-math.org
G6-M2-SE-B1-1.3.1-01.2016

This page intentionally left blank

Lesson 19: The Euclidean Algorithm as an Application of the Long Division Algorithm

Classwork

Opening Exercise

Euclid's algorithm is used to find the greatest common factor (GCF) of two whole numbers.

1. Divide the larger of the two numbers by the smaller one.

2. If there is a remainder, divide it into the divisor.

3. Continue dividing the last divisor by the last remainder until the remainder is zero.

4. The final divisor is the GCF of the original pair of numbers.

$383 \div 4 =$ $\qquad\qquad\qquad$ $432 \div 12 =$ $\qquad\qquad\qquad$ $403 \div 13 =$

Example 1: Euclid's Algorithm Conceptualized

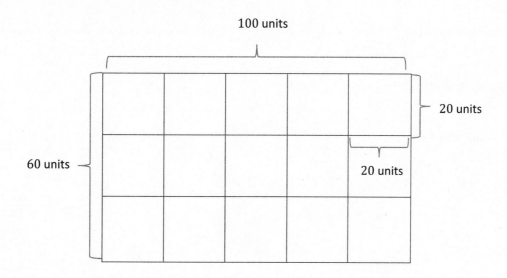

Example 2: Lesson 18 Classwork Revisited

a. Let's apply Euclid's algorithm to some of the problems from our last lesson.

   i. What is the GCF of 30 and 50?

   ii. Using Euclid's algorithm, we follow the steps that are listed in the Opening Exercise.

b. Apply Euclid's algorithm to find the GCF $(30, 45)$.

Example 3: Larger Numbers

GCF $(96, 144)$ GCF $(660, 840)$

©2015 Great Minds. eureka-math.org
G6-M2-SE-B1-1.3.1-01.2016

EUREKA
MATH™

Example 4: Area Problems

The greatest common factor has many uses. Among them, the GCF lets us find out the maximum size of squares that cover a rectangle. When we solve problems like this, we cannot have any gaps or any overlapping squares. Of course, the maximum size squares will be the minimum number of squares needed.

A rectangular computer table measures 30 inches by 50 inches. We need to cover it with square tiles. What is the side length of the largest square tile we can use to completely cover the table without overlap or gaps?

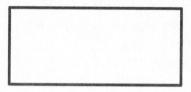

    a.   If we use squares that are 10 by 10, how many do we need?

    b.   If this were a giant chunk of cheese in a factory, would it change the thinking or the calculations we just did?

    c.   How many 10 inch × 10 inch squares of cheese could be cut from the giant 30 inch × 50 inch slab?

©2015 Great Minds. eureka-math.org
G6-M2-SE-B1-1.3.1-01.2016

Problem Set

1. Use Euclid's algorithm to find the greatest common factor of the following pairs of numbers:
    a. GCF (12, 78)
    b. GCF (18, 176)

2. Juanita and Samuel are planning a pizza party. They order a rectangular sheet pizza that measures 21 inches by 36 inches. They tell the pizza maker not to cut it because they want to cut it themselves.
    a. All pieces of pizza must be square with none left over. What is the side length of the largest square pieces into which Juanita and Samuel can cut the pizza?
    b. How many pieces of this size can be cut?

3. Shelly and Mickelle are making a quilt. They have a piece of fabric that measures 48 inches by 168 inches.
    a. All pieces of fabric must be square with none left over. What is the side length of the largest square pieces into which Shelly and Mickelle can cut the fabric?
    b. How many pieces of this size can Shelly and Mickelle cut?

©2015 Great Minds. eureka-math.org
G6-M2-SE-B1-1.3.1-01.2016